Easy as π?

Springer
New York
Berlin
Heidelberg
Barcelona
Hong Kong
London
Milan
Paris
Singapore
Tokyo

O.A. Ivanov

Easy as π?

An Introduction to Higher Mathematics

Translated by Robert G. Burns

With 60 Illustrations

 Springer

O.A. Ivanov
Department of Mathematics and
 Mechanics
St. Petersburg State University
Bibliotechnaya pl. 2, Stary Petergof
St. Petersburg, 198904
Russia

Translator:
Robert G. Burns
Department of Mathematics
 and Statistics
York University
4700 Keele Street
Toronto, Ontario MJ3 1P3
Canada

Library of Congress Cataloging-in-Publication Data
Ivanov, O.A. (Oleg A.)
  Easy as pi? : an introduction to higher mathematics / O.A. Ivanov.
     p.   cm.
  Includes bibliographical references and index.
  ISBN 0-387-98521-2 (softcover : alk. paper)
    1. Mathematics.   I. Title.
QA37.2.I87   1998
510—dc21   98-16710

Printed on acid-free paper.

First Russian Edition: Избранные главы элементарной математики, St. Petersburg, 1995.

Production managed by Timothy Taylor; manufacturing supervised by Jeffrey Taub.
Photocomposed copy prepared from the translator's LaTeX file.
Printed and bound by Edwards Brothers, Inc., Ann Arbor, MI.
Printed in the United States of America.

9 8 7 6 5 4 3 2

ISBN 0-387-98521-2 Springer-Verlag New York Berlin Heidelberg   SPIN 10773312

Foreword

The present book is rare, even unique of its kind, at least among mathematics texts published in Russian. You have before you neither a textbook nor a monograph, although these selected chapters from elementary mathematics certainly constitute a fine educational tool. It is my opinion that this is more than just another book about mathematics and the art of teaching that subject. Without considering the actual topics treated (the author himself has described these in sufficient detail in the Introduction), I shall attempt to convey a general idea of the book as a whole, and describe the impressions it makes on the reader.

Almost every chapter begins by considering well-known problems of elementary mathematics. Now, every worthwhile elementary problem has hidden behind its diverting formulation what might be called "higher mathematics," or, more simply, mathematics, and it is this that the author demonstrates to the reader in this book. It is thus to be expected that every chapter should contain subject matter that is far from elementary. The end result of reading the book is that the material treated has become for the reader "three-dimensional" as it were, as in a hologram, capable of being viewed from all sides.

It is difficult to say exactly how this effect is achieved, whether through the apt choice of problems, or their well-planned arrangement, or the judicious placement of an exercise (whose formulation is by itself often important), or even literary technique—all are significant: the subject matter, style, language, and also a sense of the author himself behind the text.

As with every serious book, far from everyone will find "Easy as π?" so easy: in order for the reader to understand it in all of its ramifications, he or she will need to be possessed already of a rather sophisticated culture—in the present case mathematical. It is even possible that the greatest pleasure from reading this book will be derived by the professional mathematician. Be that as it may, one hopes that some of its readers will at least trace through the developments out of the elementary topics, as expounded here, and then relay these to his or her students. In any case it is certainly a good thing that such a book has been written, and I recommend it with pleasure to a wide readership.

A.S. Merkurjev
University of California at Los Angeles

(Formerly at St. Petersburg State University, Russia)

Thus ev'ry kind their pleasure find.
Robert Burns (1759–1796)

Preface

This book represents an expanded version of lecture notes of a course given by the author over a period of several years to fourth- and fifth-year students specializing in education in the Mathematics–Mechanics Faculty of St. Petersburg (formerly Leningrad) University. This course (consisting of 60 hours' worth of lectures) was conceived as being of a summarizing, general–mathematical nature, intended to make extensive reference to the concepts and propositions of the more basic mathematics courses. It had seemed clear that mathematics teachers in particular would like the idea of such a course, with "mathematics as a whole" as its subject–matter, i.e., without the traditional subdivision into algebra, analysis, geometry, etc. It goes without saying that this course was not in the least intended to replace the more standard basic courses, but rather to supplement and clarify them. Thus *Easy as π?* should be suitable as a text for a wide readership interested in being introduced to modern higher mathematics, i.e. in "enriching" their knowledge of elementary mathematics.

However, since it was originally conceived as a textbook for teachers at (specialized) academic middle schools and highschools, and for those training to become such teachers, a few words relevant to this aspect of the book may be in order. (Note in this connection that almost every chapter concludes with pedagogical remarks concerning the material of that chapter.) A teacher at such an institution should have sufficient depth and breadth of knowledge and understanding of mathematics to be capable of designing a mathematical curriculum appropriate to the mandate of his or her institution. For this reason in particular, the education of a teacher must not be restricted to the immediate material being taught. To give an example: while it is not appropriate for a teacher even to mention Peano's axioms for arithmetic in class (except perhaps where the class consists of an "elite," in the best sense of that word), he or she should have an understanding of these axioms so as to be capable at least of explaining to an intelligent eighth–grader where the multiplication table comes from, or helping a clever eleventh–grader for whom the method of mathematical induction is not obvious.

I am grateful to the translator, Robert Burns, not only for his expert translation of this book, but also for his editorial work. I am sure that the book has been considerably improved as a result of the continual communication between author and translator throughout the course of the translation.

O.A. Ivanov
St. Petersburg, Russia

Translator's Acknowledgments

I am grateful to Lydia Burns, Xiao-Min Dong, Anna and Wolfgang Herfort, Siu-Man Kam, Yuri Medvedev, Abe Shenitzer, and Pavel Zalesskii variously for help with Russian, mathematics, and LaTeX, and for many other kindnesses, while this translation was being carried out at York University, Toronto, and the Technical University, Vienna.

Robert G. Burns
Toronto, Canada

Contents

Introduction

The material expounded in this book is divided up by theme or topic, with adjoining sections related for the most part by the internal structure of their expositions. Most chapters begin with problems of elementary mathematics which might serve as a basis for the work of a school "mathematics circle" (i.e. an after-school mathematics session for interested students) or an optional course. These are followed by concepts of "higher" mathematics related to the elementary problems by analogies of method or idea, or else by foundational questions concerning these. Examples showing how this is done will be given below. The "higher" mathematics appearing in this book consists of classical results with instructive but not unduly complicated proofs, always expounded from a unified algebraic-geometric point of view, but on occasion involving concepts which at first glance may seem far removed from the elementary problems serving as their source. It should be emphasized also that among these more advanced concepts and results there are some that are not treated in the traditional core of applied or pure mathematics university courses. At the end of each chapter there are appended remarks of a methodological character. In view of the fact that the author has consciously aimed at conciseness—the book is after all a summary!—only hints at the solution are given to some of the problems, and in certain arguments some of the steps are left as exercises for the reader. Thus between the formulation of a result and the symbol □ indicating the end of the proof, there may well appear just a sequence of exercises.

Bibliographical references are given only in those cases where the cited source has been used in an essential way in the text. The author should also confess that the content and style of exposition of the book reflect the pedagogical ideas of F. Klein [17] and G. Pólya [26]—at least as he understands them.

We point out once again that every chapter except 8–10 contains elementary problems. The problems in Chapter 9, although they cannot be considered elementary, nonetheless relate to the concept of the derivative as taught in high school.

The following detailed description of the contents of the book covers mainly the development of the nonelementary topics.

In Chapter 1, " Induction," the basic mathematical object is the set $\mathbb{N}$ of natural numbers. In attempting to prove the principle of mathematical induction one is inevitably led to the necessity of giving a formal description of this set. The logical simplicity of the axioms suggested by Peano allows one to trace step by step how $\mathbb{N}$ acquires its algebraic (or rather arithmetic) structure. (In particular, the reader is asked by way of an exercise to prove(!) that $2 \times 2 = 4$.) The difference between the formal application of mathematical induction and inductive reasoning (not scientific induction) is stressed (see Problem 2), and the reader's attention is directed to the often implicit use of the latter. As a rather separate topic the question of the definition of the number of elements of a set is also considered. One of the exercises asks for a proof of the (obvious) formula $|A \times B| = |A||B|$. This chapter also includes a construction of the set $\mathbb{Z}$ of integers.

Chapter 2, "Combinatorics," is devoted mainly to motivating, introducing, and applying the functional approach to solving combinatorial problems, i.e., to the method of "generating functions". The method is motivated by means of an example illustrating Newton's binomial theorem. Prior to this the binomial coefficients are introduced in three different, though equivalent, ways, in the most geometric of which the basic recurrence relation between the binomial coefficients becomes practically obvious. Next, using as example the derivation of the explicit formula for the Fibonacci numbers, it is shown that in order to understand fully the method of generating functions, it is appropriate to consider the ring of formal power series. The main combinatorial problem treated in this chapter is that concerning the number of partitions of a natural number. In addition, a simpler problem on the enumeration of graphs is considered, and the chapter concludes with an analytic proof of Stirling's asymptotic formula for $n!$.

The chief preoccupation in Chapter 3, "Geometric Transformations," is with the algebraic (group) structure on the set of Euclidean motions of the plane. By way of motivating the basic definition, several elementary problems are considered (the most important of which are Nos. 7 and 8), and its most significant applications here is the algebraic classification of the periodic patterns along a band, or frieze patterns (or "ornaments"). Of particular methodological importance in this chapter is the calculation using coordinates, of the composite of two transformations.

The exposition of the material of Chapter 4, "Inequalities," essentially follows that of the classical treatise [12] of Hardy, Littlewood, and Pólya. In particular, five proofs are given of the classical inequality between the arithmetic and geometric means due to Cauchy. In distinction to [12] however, more attention is paid to the geometrical interpretation of inequalities; thus, for instance, the Minkowski norm is introduced, and the theorem on the equivalence of norms on a finite-dimensional vector space is proved. The coefficients in a Fourier series are also interpreted geometrically using in the associated argument Wirtinger's inequality. The latter inequality is also used to solve the isoperimetric problem. (That argument also employs a special case of Green's theorem, expressing the area of a planar region in terms of a line integral around its boundary.)

Chapter 5 contains a great variety of material, although its title, "Sets, Equations, and Polynomials," reflects fairly accurately the underlying theme: the investigation of properties of sets of points via systems of equations defining them, i.e., algebraic geometry. We mention two topics. The first of these, although very straightforward, is educationally useful: the representation of sets given by very simple equations (or inequalities). As the source for the second topic we take the problem of constructing a polynomial with integer coefficients having $\sqrt{2} + \sqrt{3}$ as one of its roots. The standard solution of this exploits the idea of eliminating an unknown from a system of equations. The further development of this idea—the method of elimination—is applied in the proof of Bézout's theorem on the number of points in the intersection of two plane algebraic curves. Then, by means of a seemingly slight reformulation of the problem of finding such a polynomial, the reader is introduced to the concept of a finite field extension.

The theme of Chapter 6, "Graphs," is standard fare in expositions of elementary mathematics. Also standard in this respect is a proof of Euler's formula for a planar graph. However, then the question as to why there is no immediate analogue of that formula for arbitrary graphs on the torus leads to the necessity of formulating and then proving the Jordan curve theorem (in the case where the continuous closed simple plane curve is piecewise linear, i.e., polygonal). The proof of this version of the theorem given here uses in particular the general mathematical concept of the index of a point relative to a plane curve (actually, in our situation relative to a simple plane polygonal arc). By means of the notion of a "pairing" in a bipartite graph, a geometrically transparent proof is obtained of the Bernstein–Cantor–Schroeder theorem on the comparison of "sizes" of arbitrary sets.

The idea behind most of the material considered in Chapter 7, "The Pigeonhole Principle," is very simple: if more than n objects (e.g., pigeons) are distributed among n boxes, then some box will contain at least two of the objects. (This principle is sometimes attributed to Dirichlet.) This idea in various guises provides the key to the solution of many Mathematical Olympiad problems, and is also instrumental, as is shown in this chapter, in answering the following question: When is -1 a quadratic residue with respect to a given modulus? (These results are then used as lemmas in proofs of the theorems of Fermat–Euler and Lagrange on the representability of a natural number as a sum of two and four squares of integers, respectively.) The most interesting results are obtained when the pigeonhole principle is applied in the situation of infinite sets, with measure (length, area, or volume) replacing the number of elements. The abstract results of this kind included in this chapter are Poincaré's "recurrence theorem" for measure-preserving transformations and Minkowski's lemma concerning centrally symmetric convex subsets in $\mathbb{R}^n$ (providing a sufficient condition for such a subset to contain a point of a given lattice other than the origin). Each of these results has a great many applications, including some sufficiently elementary ones. In particular, Minkowski's lemma is crucial in the proofs we give of the above-mentioned theorems of Fermat–Euler and Lagrange, and from the theorem of Poincaré a well-known paradoxical assertion of physics (concerning "recurrence") follows. In this chapter various familiar concepts and results arise naturally: Lagrange's theorem on the index of a subgroup of a finite group; variational systems and the differentiability with respect to the initial conditions of a system of differential equations; Fubini's theorem; and closed subgroups.

The central topic of Chapter 8, "The Quaternions," might seem at first glance not to accord with the basic concept of the book. As justification for including them the author first demonstrates how they afford a transparent interpretation of the Eulerian identity appearing at the end of the preceding chapter. Following this the algebraic closure of the field $\mathbb{C}$, the field of complex numbers, i.e. the nonexistence of a number field properly containing $\mathbb{C}$, is considered. The study of the properties of the skew-field of quaternions leads naturally to the introduction of a new algebraic structure, that of an "algebra," important examples of which are furnished by spaces of matrices with their natural operations. As a further application, quaternions are used to describe the rotations of the Euclidean spaces $\mathbb{R}^3$ and $\mathbb{R}^4$. Another result considered in this chapter is Stiefel's theorem: If a multiplication can be defined on the vector space $\mathbb{R}^n$ making it a division algebra, then on the unit sphere in $\mathbb{R}^n$ there exist $n - 1$ linearly independent vector fields. A proof (not quite complete) is also given of the classical theorem of Frobenius classifying the finite-dimensional division algebras over $\mathbb{R}$.

The material of Chapter 9, "The Derivative," consists essentially of applications of that concept (for the most part to problems of geometry and mechanics, for instance to determining the position of a point the sum of whose distances from three given fixed points is least (Problem 3)), based on the classical theorem of differential calculus (of Fermat), and the formula for the derivative of the length of a vector–function. The problem of resolving the acceleration vector of a mass-point into components parallel and perpendicular to a prescribed trajectory, leads to a generalization of the formula for centripetal acceleration $w_N = v^2/R$, familiar from highschool physics courses; in our more general situation, R is the radius of curvature of the given curve. From Newton's universal law of gravitation follow, via calculus and certain types of constancy exhibited by associated vector–functions, Kepler's laws governing the motion of a body in a gravitational field. A separate topic concerns the equation for the motion of a pendulum, a particular case of a one-dimensional conservative system, i.e., a system possessing a first integral (the energy). The stability of the stationary points of such a system is investigated, and the solution of the system of first approximation (i.e., the linear approximation to the given system). Note also the interpretation given in this chapter of two standard definitions of the number e by means of approximate solutions of differential equations.

Problem 10, which should be fully accessible to students, perhaps even solvable independently by some of them, introduces an important topic of Chapter 9. The simple and transparent solution of this problem does not immediately generalize to higher dimensions. In this connection a method of solution of equations by iteration is expounded, having as its theoretical basis Banach's theorem on contracting maps. This is then used to establish the existence of solutions of a certain type of differential equation and to prove the inverse function theorem. A particular case of the Morse–Sard theorem is also proved, whence follows, in particular, the existence of a planar fourth-degree curve which is the union of four "ovals."

It should be noted that notwithstanding its apparent difficulty, practically all of the material of Chapter 9 is within the scope of specialized or optional courses such as might be offered at a highschool specializing in mathematics and science.

A special place in this book is occupied by Chapter 10, "The Foundations of Analysis," where the main goal is to give a precise definition of, and subject to scrutiny from all sides, the fundamental mathematical object *par excellence*, the basis in particular for mathematical analysis, namely the set $\mathbb{R}$ of real numbers. We introduce Dedekind's model of the real numbers (as "cuts" in the set $\mathbb{Q}$ of rationals) and Cauchy's model (where the reals are defined as equivalence classes of Cauchy sequences of rationals). It is interesting to observe that the axiom of continuity is more easily established using the first of these models, while the second lends itself more readily to the introduction of the arithmetic operations. The second is the more general in the sense that the process of (Cauchy) completion can be applied in any metric space, in particular to $\mathbb{Q}$ endowed with the p-adic metric. In the last section of the chapter it is shown that every nontrivial norm on the field of rational numbers is equivalent to either the standard one or the p-adic one for some p.

Another mathematical object described in this chapter, one rarely encountered in standard courses in analysis, is the "nonstandard line," an ordered field containing $\mathbb{R}$ and nonzero infinitesimal elements. The construction of a model of this line is to some extent analogous to the construction of the Cauchy model of $\mathbb{R}$. The link between standard and "nonstandard" analysis is provided by the "translation principle," the essence of which is that every true assertion of standard analysis remains true when the field $\mathbb{R}$ is replaced wherever it occurs in the assertion by the nonstandard number line. The exhibited examples make it evident that theorems of ordinary analysis become easier to prove in the context of nonstandard analysis, so that the translation principle is crucial in the construction of the latter theory. Without going into detail, we remark that in our construction of a model of the nonstandard line we find that we need to establish the existence of a family of subsets of $\mathbb{N}$ with highly nontrivial properties. Moreover, this family is "nonconstructive," since the proof depends in an essential way on Zorn's lemma.

How should this book be read? It will be accessible *in toto* to the reader familiar with the basic concepts and results of algebra, analysis, geometry, topology, and differential equations as these topics are covered in first- and second-year university mathematics courses. However, it is likely that reading it without pen and paper to hand will be possible only for the professional mathematician, interested perhaps in the style or the interpretations of various of the results or themes. For others, reading this book will not be a very simple matter.

The author has striven as far as possible to ease the reader's task, organizing the material in such a way that he or she may plunge in at almost any point, since all of the chapters and a majority even of the sections are practically independent. A large portion of the book should be accessible to students of the higher grades (provided, of course, that they are prepared to make a sufficient effort). As for those for whom the author originally intended this book, it is his most earnest wish that it help them to understand better the subject they teach (or will teach) and also how it is situated within that region of human culture called "mathematics."

It is the author's hope that this book will be useful more generally to all who study or teach mathematics, and also that it conforms to the spirit and tradition of the Leningrad school of mathematics education. It could not have been written without the author's having had the happy experience of studying first at the Leningrad School for Physics and Mathematics (now Gymnasium) No. 30 under the brilliant teachers Anatolii Anatolyevich Vanyeev and Yosef Yakovlevich Verebeĭchik, and then in the Mathematics–Mechanics Faculty of Leningrad University under the supervision of such outstanding mathematicians and extraordinary teachers as Viktor Aleksandrovich Pliss and Vladimir Abramovich Rokhlin.

I am deeply grateful to Tatyana Mikhaĭlovna Mishchenko for her constant encouragement and for many discussions concerning the content of the book, to Aleksandr Sergeevich Merkurjev for taking on the task of reading the manuscript, and to Irina Nikolaevna Ryazanova for her editorial work. I also thank all those who taught me the language TEX.

Lastly, I must acknowledge the part played by my family, in particular their forbearance over the period when I was writing, rewriting, and preparing for publication this introduction to higher mathematics, which I hereby dedicate to *Tanya, Tanya*, and *Iana*.

O.A. Ivanov
St. Petersburg, Russia

CHAPTER 1

Induction

1.1 Principle or method?

The phrases "the method of mathematical induction," "the principle of mathematical induction," and "an inductive argument" are sometimes used synonymously. One of the goals of the present section is to highlight their intrinsic differences of meaning.

Problem 1. Prove the identity $\sum_{k=1}^{n} k^3 = \frac{n^2(n+1)^2}{4}$.

Here is a proof using the "method of mathematical induction": A direct, easy check shows that the identity holds in the case $n = 1$. Next, from the assumption that $\sum_{k=1}^{n} k^3 = n^2(n+1)^2/4$, we infer that $\sum_{k=1}^{n+1} k^3 = (n+1)^2(n+2)^2/4$ as follows:

$$\sum_{k=1}^{n+1} k^3 = \sum_{k=1}^{n} k^3 + (n+1)^3 = \frac{n^2(n+1)^2}{4} + (n+1)^3$$

$$= \frac{(n+1)^2}{4}(n^2 + 4n + 4) = \frac{(n+1)^2(n+2)^2}{4}.$$

Hence the given identity is valid for all natural numbers n.

Note that we might have, so to speak, "avoided" induction by first observing that $k^3 = \frac{k^2(k+1)^2}{4} - \frac{(k-1)^2 k^2}{4}$ and then adding the equations $1^3 = \frac{1^2 \cdot 2^2}{4} - 0$, $2^3 = \frac{2^2 \cdot 3^2}{4} - \frac{1^2 \cdot 2^2}{4}, \ldots, n^3 = \frac{n^2(n+1)^2}{4} - \frac{(n-1)^2 n^2}{4}$.

The above proof illustrates the use of the method of mathematical induction as a convenient means for proving statements already formulated.

Problem 2 (The Tower of Hanoi). On one of three vertical pegs there are threaded 10 disks of different diameters (each with a hole at its center so that it can be slid on and off the pegs) arranged in order of increasing diameter, so that each disk rests on one larger than itself. What is the least number of moves of one disk at a time from one peg to another by means of which all the disks can be transferred to one of the other two pegs, without a disk ever being placed on top of one smaller than it?

We shall solve this problem immediately for an arbitrary number n of disks. On investigating the cases $n = 1, 2, 3$, we find that $1, 3$, and 7 moves respectively are required. In order to transfer a tower of four disks, it is clear that first the upper three must be moved to the third peg (7 moves), then the bottom, largest, disk to the free second peg (1 move), and finally the three smaller disks placed on top of the latter (7 moves), making altogether $7 + 1 + 7 = 15$ moves. From this argument it is clear that in general we have the recurrence relation $p_{n+1} = 2p_n + 1$, where p_n denotes the least number of moves for the tower of n disks. Using this formula to calculate the next few values of p_n, we obtain $31, 63, 127, \ldots$, whence one guesses that in fact $p_n = 2^n - 1$. This formula is then easily proved "by induction."

"The Tower of Hanoi" is one of the best problems on induction, yet the "method of mathematical induction" itself plays no very large role in its solution. Close examination of the above solution reveals that its essence lies in the construction of a recursive algorithm. This can be more clearly brought out as follows: Let $\mathcal{T}_n(i, j)$ denote the sequence of moves needed to transfer n disks from the ith peg of the three to the jth. Then $\mathcal{T}_1(i, j) = \{$removal of a disk (the top one) from the ith peg to the jth$\}$, and

$$\mathcal{T}_{n+1}(i, j) = \{\mathcal{T}_n(i, k), \mathcal{T}_1(i, j), \mathcal{T}_n(k, j)\}.$$

Here are two more good problems of this general type.

Problem 3. Prove that a checkerboard with $2^n \times 2^n$ squares (or "cells") from which one square has been removed can be covered exactly by "triominoes" ("corner-pieces") of the form

(If we subdivide the checkerboard into four large squares (or "quadrants"), each consisting of $2^{n-1} \times 2^{n-1}$ cells, and place a single corner-piece so that it covers exactly one cell of each of the three quadrants other than the one with the missing cell, then we are in the situation of four checkerboards of smaller size, from each of which a cell has been removed.)

Problem 4. Into how many parts do (a) n points subdivide a line; (b) n straight lines subdivide the plane, if no two of the lines are parallel and no three are concurrent, i.e., meet in a single point?

That a proof "by induction" qualifies as a genuine proof might be considered self-evident. A sceptic might be convinced by the following argument: Suppose we have a statement involving the variable n which is true for $n = 1$ and is such that whenever it is true for some value k of n, then it is true for $n = k + 1$, yet is *not* true for all natural numbers n. Let n_0 denote the smallest number for which the statement is false. Then clearly $n_0 > 1$, so that the integer $n_0 - 1$ is a natural number, and furthermore, one for which the statement is true. However, then it must be true also for n_0!

A greater sceptic might then raise the following question: Why must every nonempty subset of N have a least element? After all, Q has nonempty subsets without least elements. Let $E \subseteq N$, and let $a \in E$. If a is not least in E, then there is an element $a_1 \in E$ such that $a_1 < a$. If a_1 is not least in E, then we can find a yet smaller number $a_2 \in E$, and so on, this process ultimately terminating, since we cannot go below 1, the least of all natural numbers. But how can we avoid this "and so on"?

1.2   The set of integers

Suppose for the time being that the set N of natural numbers exists (whatever that may mean) and that the usual operations of addition and multiplication, and also the natural ordering, are given on N. We shall now construct from N the set Z of integers, or whole numbers.

First write $Z_+ = N \cup \{0\}$, where 0 is any object not contained in N. We extend the arithmetic operations to embrace this new element, by defining $a + 0 = 0 + a = a$, $a \cdot 0 = 0 \cdot a = 0$ for every $a \in Z_+$, and set $0 < a$ for every $a \in N$.

Exercise. Show that in fact the property $a \cdot 0 = 0 \cdot a = 0$ is a consequence of the distributive law and the definition of zero as the identity element with respect to addition.

Now consider the set Z of ordered pairs (a, b), $a, b \in Z_+$, and the relation $\sim$ on this set defined by $(a, b) \sim (a', b')$ if $a + b' = a' + b$. Since this is an equivalence relation, i.e. is reflexive: $(a, b) \sim (a, b)$; symmetric: $(a, b) \sim (a', b')$ implies $(a', b') \sim (a, b)$; and transitive: $(a, b) \sim (a', b')$ and $(a', b') \sim (a'', b'')$ imply $(a, b) \sim (a'', b'')$ (verify these!), the set Z is partitioned into pairwise disjoint "equivalence classes," i.e., maximal subsets with respect to the property that every two pairs in a subset be equivalent. We shall denote the class containing a pair (a, b) by $[a, b]$:

$$[a, b] := \{(a', b') \in Z | (a', b') \sim (a, b)\}.$$

We now define Z as the set of all of these classes[1] (a "factor set" of Z).

---

[1] *Translator's note.* An alternative, rather more obvious and direct, way of constructing the integers from the natural numbers is simply to supplement $N \cup 0$ with further symbols $-n$, $n \in N$, representing the negative integers, and extend addition and multiplication to this enlarged set in the familiar way. (One has then, of course, to check that the usual rules of arithmetic—i.e., defining conditions of a commutative ring—continue to hold.) This

We now define the arithmetic operations on $\mathbb{Z}$, beginning with *multiplication*. For pairs $(a, b), (c, d) \in \mathcal{Z}$ we set $(a, b) \cdot (c, d) := (ac + bd, ad + bc)$. Then if $(a, b) \sim (a', b')$, we have

$$ac + bd + a'd + b'c = (a + b')c + (a' + b)d$$
$$= (a' + b)c + (a + b')d = ad + bc + a'c + b'd,$$

which shows that $(a, b) \cdot (c, d) \sim (a', b') \cdot (c, d)$. Now, for any $\alpha, \beta \in \mathbb{Z}$, choose $(a, b), (c, d) \in \mathcal{Z}$ such that $[a, b] = \alpha$, $[c, d] = \beta$, and define

$$\alpha \cdot \beta := [(a, b) \cdot (c, d)] = [ac + bd, ad + bc].$$

The above argument shows that the class on the right-hand side of this equation is independent of the choice of the pairs (a, b) and (c, d) from the classes α and β respectively. For instance, $[0, 2] \cdot [1, 0] = [0, 2] = [9, 11] = [(1, 3) \cdot (3, 2)]$ (see the diagram). Next we define *addition* by $[a, b] + [c, d] = [a + c, b + d]$, and $[a, b] < [c, d]$ to mean $a + d < b + c$. Finally, if we identify each number $a \in \mathbb{Z}_+$ with the class $[a, 0] \in \mathbb{Z}$, then we have $\mathbb{Z}_+ \subset \mathbb{Z}$. (Why is this an embedding?)

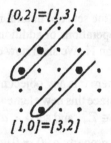

Exercise. Verify the usual properties of the arithmetic operations and the order relation on the set $\mathbb{Z}$ of integers.

Writing $-1 := [0, 1]$, we have $[a, b] = [a, 0] + [0, b] = [a, 0] + [0, 1] \times [b, 0] = a + (-1)b$, whence we see that $\mathbb{Z}$ is in effect just the set of formal differences $a - b$, $a, b \in \mathbb{Z}_+$. The source of the seemingly obscure rule for multiplying pairs (a, b) now becomes evident.

Thus starting with the natural numbers we have constructed the integers; but on what can we base the definition of the natural numbers? Since we have to stop somewhere, we might as well resort at this point to the axiomatic method of simply taking undefined objects and relations between them, and postulating rules (or "axioms") that these are required to obey (much as in Euclidean geometry). We shall formulate these axioms in set-theoretical terms, although this is not essential;

---

amounts to choosing the pairs of the form $(0, n)$ as representatives of the above equivalence classes. The approach in the text has the advantage of being consistent with the way negative numbers arise in practice often as differences, e.g., when expenditure exceeds income, and also affording a good introduction to the ideas of equivalence, equivalence class, and "congruences on semigroups."

embedded in the context of set theory the axioms take on the aspect merely of defining conditions.

1.3 Peano's axioms

We take the natural numbers to be the elements of any set $\mathcal{N}$ that comes with a "distinguished" element, denoted by 1, and a mapping $s : \mathcal{N} \to \mathcal{N}$ (so that the natural numbers are defined by the triple $\{\mathcal{N}, 1, s\}$) with the following properties [4]:

P1. The map s should be injective; i.e., if $x, y \in \mathcal{N}$ and $x \neq y$, then $s(x) \neq s(y)$;
P2. $1 \notin s(\mathcal{N})$, i.e. the distinguished element should not be the image of any element of $\mathcal{N}$;
P3. If $M \subseteq \mathcal{N}$ has the properties that $1 \in M$ and $s(x) \in M$ for all $x \in M$, then $M = \mathcal{N}$.

We shall call a subset M of the set $\mathcal{N}$ *inductive* if $1 \in M$ and $s(x) \in M$ for all $x \in M$. The last axiom may then be reformulated thus: $\mathcal{N}$ contains no proper, i.e., different from itself, inductive subsets. (The above axiom system was introduced in the 1890s by the Italian mathematician Giuseppe Peano.) The role one has in mind for $s(x)$ is that of the "successor" of x, i.e., the natural number $x + 1$ succeeding it; however, addition on the set $\mathcal{N}$ has yet to be defined. Peano's axioms form a "complete" axiom system in the sense that any two triples $\{\mathcal{N}, 1, s\}$, $\{\mathcal{N}', 1', s'\}$ satisfying the axioms are isomorphic, i.e., there is a one-to-one correspondence (bijection) $\Phi : \mathcal{N} \to \mathcal{N}'$ such that $\Phi(s(x)) = s'(\Phi(x))$. (We omit the proof of this.)

We now deduce some consequences of Peano's axioms.

Lemma 1. *One has* $s(\mathcal{N}) = \mathcal{N} \setminus \{1\}$ *and* $s(a) \neq a$ *for all* $a \in \mathcal{N}$.

(Prove that the sets $M_1 = \{1\} \cup s(\mathcal{N})$ and $M_2 = \{a \in \mathcal{N} | s(a) \neq a\}$ are inductive.) $\square$

A (*binary*) *operation* on a set is given by a map F associating with each ordered pair (a, b) of elements an element $c = F(a, b)$ of the set. Commutativity of an operation is equivalent to the map F being symmetric, and associativity to the following condition on F:

$$F(a, F(b, c)) = F(F(a, b), c) \ \forall a, b, c.$$

Consider the following two properties of an operation on $\mathcal{N}$ given by a map F:

A1. $F(a, 1) = s(a)$ for all $a, b \in \mathcal{N}$.
A2. $F(a, s(b)) = s(F(a, b))$ for all $a, b \in \mathcal{N}$.

Theorem 1. *An operation* $F : \mathcal{N} \times \mathcal{N} \to \mathcal{N}$ *having the two properties A1 and A2 is commutative and associative.*

Associativity is established as follows: For an arbitrary fixed pair of elements $a, b \in \mathcal{N}$, set

$$M := \{c \in \mathcal{N} \mid F(a, F(b, c)) = F(F(a, b), c)\}.$$

We have $F(a, F(b, 1)) = F(a, s(b))$ (by property A1) $= s(F(a, b))$ (by property A2) $= F(F(a, b), 1))$ (again by property A1), whence $1 \in M$. Now let c be any element of M. Then $F(a, F(b, s(c))) = F(a, s(F(b, c))) = s(F(a, F(b, c))) = s(F(F(a, b), c))$ (here we are using the fact that $c \in M$) $= F(F(a, b), s(c))$, whence $s(c) \in M$. This shows that M is an inductive subset of $\mathcal{N}$, whence $M = \mathcal{N}$.

Exercise. Complete the proof by establishing commutativity.

(The set $\{a \in \mathcal{N} \mid F(a, 1) = F(1, a)\}$ is inductive.) $\square$

Theorem 2. *There is at most one map F with properties A1 and A2.*

Let F and F' be two such maps, and taking an arbitrary fixed element $a \in \mathcal{N}$, consider the set

$$M := \{b \in \mathcal{N} \mid F(a, b) = F'(a, b)\}.$$

Since $F(a, 1) = s(a) = F'(a, 1)$, we have $1 \in M$. For any element $b \in M$ we have $F(a, s(b)) = s(F(a, b)) = s(F'(a, b)) = F'(a, s(b))$, whence $s(b) \in M$. Hence M is an inductive set, and therefore $M = \mathcal{N}$, which implies that $F = F'$. $\square$

1.4   Addition, order, and multiplication

Consider the subset M of $\mathcal{N}$ consisting of just those elements $a \in \mathcal{N}$ for which there exists a map $f_a : \mathcal{N} \to \mathcal{N}$ such that $f_a(1) = s(a)$ and $f_a(s(b)) = s(f_a(b))$ for all $b \in \mathcal{N}$. Defining $f_1(b) := s(b)$, we have $f_1(1) = s(1)$ and $f_1(s(b)) = s(s(b)) = s(f_1(b))$, whence we see that $1 \in M$. For each $a \in M$, we define $f_{s(a)}$ via the formula $f_{s(a)}(b) := s(f_a(b))$. One then has $f_{s(a)}(1) = s(f_a(1)) = s(s(a))$ and $f_{s(a)}(s(b)) = s(f_a(s(b))) = s(s(f_a(b))) = s(f_{s(a)}(b))$ for all $b \in \mathcal{N}$. This shows that the set M is inductive, whence $M = \mathcal{N}$, and the formula $F(a, b) := f_a(b)$ defines a map F with properties A1 and A2. Combining this with Theorems 1 and 2, we have the following result:

Theorem 3. *There exists an operation of addition on the set $\mathcal{N}$; i.e., there exists a map F with properties A1 and A2. Furthermore, this operation (or, equivalently, the map) is unique, and is commutative and associative.*

Lemma 2. *For all $a, b \in \mathcal{N}$, we have $a + b \neq a$.*

To see this, observe first that $1 + b = b + 1 = s(b) \neq 1$ in view of Lemma 1. Further, if an element $a \in \mathcal{N}$ is such that $a + b \neq a$ for all $b \in \mathcal{N}$, then since the map s is injective, we shall have $s(a) \neq s(a+b) = s(b+a) = b+s(a) = s(a)+b$. This shows that the set $M = \{a \in \mathcal{N} \mid a+b \neq a \; \forall b \in \mathcal{N}\}$ is inductive, whence the lemma. $\square$

The order relation on the set $\mathcal{N}$ is defined as follows: we set $a > b$ if $a = b+k$ for some element $k \in \mathcal{N}$.

Theorem 4. *For any two elements $a, b \in \mathcal{N}$ exactly one of the following possibilities occurs: $a > b$; $a = b$; $a < b$.*

That no two of these possibilities can be realized for a single pair a, b follows easily from Lemma 2. To show that at least one of them is always realized, fix on an arbitrary element $b \in \mathcal{N}$ and consider the set $M := \{a \mid a > b \text{ or } a = b \text{ or } a < b\}$.

Exercise. Prove that this set M is inductive. $\square$

Corollary 1. *For every $a \in \mathcal{N}$ the elements $a, a+1$ are adjacent with respect to the above-defined ordering of $\mathcal{N}$, i.e. there is no element $b \in \mathcal{N}$ such that $a < b < a+1$.*

Corollary 2. *For every $a \in \mathcal{N}$ we have $a \geq 1$.*

Exercise. Prove these corollaries.

Theorem 5. *Every nonempty subset of $\mathcal{N}$ has a smallest element.*

Let E be any nonempty subset of $\mathcal{N}$, and consider the set $M := \{a \in \mathcal{N} \mid a \leq b \; \forall b \in E\}$, consisting of all lower bounds of E. Since the set E is nonempty, we have by Corollary 2 above that $1 \in M$, whence it follows that there is an element $u \in \mathcal{N}$ such that $u \in M$ but $u+1 \notin M$. If we can show that $u \in E$, it will follow that u is the desired least element of that set. Now, since $u+1 \notin M$, there must be an element $b_0 \in E$ such that $b_0 < u+1$. If u were not in E, we should have $u < b$ for all $b \in E$, whence in particular, $u < b_0 < u+1$, contradicting Corollary 1 above. $\square$

Multiplication on the set of natural numbers is defined analogously. One has the following theorem.

Theorem 6. *There exists precisely one map $G : \mathcal{N} \times \mathcal{N} \to \mathcal{N}$ satisfying:*

M1. $G(a, 1) = a$;

M2. $G(a, s(b)) = G(a, b) + a$.

The operation on $\mathcal{N}$ defined by the map G is commutative, associative, and distributive over addition, i.e., for all $a, b, c \in \mathcal{N}$ the equation $G(a + b, c) = G(a, c) + G(b, c)$ holds.

Exercise. Prove this theorem. $\square$

Exercise. Prove that $2 \times 2 = 4$. (Here $2 := s(1)$, $3 := s(2)$, $4 := s(3)$.)

Exercise. Prove that for all $a, b, c \in \mathcal{N}$, if $a < b$ then $ac < bc$.

1.5   The method of mathematical induction

That the method of mathematical induction is actually valid is, as we have seen, essentially axiomatic; it is one of the defining properties, assumed axiomatically to be viable, of the set of natural numbers. Hence it is more appropriate to speak not of the "method" but rather of the "principle" of mathematical induction. Here is a useful variant of that principle:

Theorem 7. *Let* $\{P(n)\}_{n \in \mathcal{N}}$ *be a sequence of mathematical propositions (statements). Suppose that for some natural number* l *the statement* $P(l)$ *is true, and that for every natural number* $n \geq l$, *if* $P(k)$ *is true for all* k *such that* $l \leq k \leq n$, *then* $P(n + 1)$ *is true. Then* $P(n)$ *is true for all* $n \geq l$.

To prove this theorem consider the natural number

$$n_0 := \min\{n \in \mathcal{N} \mid n \geq l \text{ and } P(n) \text{ false }\}.$$

Since $n_0 \geq l$ and $P(n_0)$ is false, we must have in fact $n_0 > l$. In view of the minimality of n_0 the statement $\mathcal{N}(k)$ must be true for $k = l, l + 1, \ldots, n_0 - 1$. However, then by the hypothesis of the theorem, $P(n_0)$ must also be true. This contradiction yields the desired conclusion. □

One must bear in mind that in the solution of elementary problems, especially those intended for younger students, the principle of mathematical induction is often only implicit.

Problem 5. Show that a square can be subdivided into 6, 8, or 9 smaller squares. For which other numbers can this be done?

Answer. For any number other than 2, 3, 5.

It is not difficult to subdivide a square into 6, 8, or 9 smaller squares, and subdividing it into 7 such is easy: one just draws a big "+" sign in the given square, and then a smaller one in one of the resulting quarters (see the diagram). The latter subdivision suggests an idea for a solution: if the square has a subdivision into k smaller squares, then it can be subdivided into $k + 3$ squares. Certainly for younger students the argument should end at this point; but of course there is an induction involved, and a rigorous proof would depend on the principle of mathematical induction, in fact in the form of the above variant of that principle.

Problem 6. Prove that any sum of more than 7 cents can be made up out of 3- and 5-cent coins.

There are problems that do not readily yield to a frontal assault using induction, whereas a stronger assertion yields easily.

Problem 7. Prove the inequality $\frac{1}{n+1} + \frac{1}{n+2} + \cdots + \frac{1}{2n} < \frac{3}{4}$.

Denoting by a_n the left-hand side of this inequality, we have that

$$a_{n+1} = a_n + \frac{1}{2n+1} + \frac{1}{2n+2} - \frac{1}{n+1} = a_n + \frac{1}{2n+1} - \frac{1}{2n+2} > a_n,$$

so that it is not clear from this how one can infer from $a_n < \frac{3}{4}$ that $a_{n+1} < \frac{3}{4}$. However, it *does* follow without difficulty that if $a_n \leq \frac{3}{4} - \frac{1}{4n}$, then $a_{n+1} \leq \frac{3}{4} - \frac{1}{4n+4}$ (verify!), whence it follows in turn, once the case $n = 1$ has been checked, that for all natural numbers n, one has $a_n \leq \frac{3}{4} - \frac{1}{4n} < \frac{3}{4}$.

As formulated, Problem 7 contains an ellipsis $(\ldots)$, contrary to our wish expressed at the outset to avoid the phrase "and so on." This can be avoided as follows: In view of the order relation on $\mathbb{N}$, we may for each $k \in \mathbb{N}$ consider the subset $I_k := \{n \in \mathbb{N} \mid n \leq k\}$. The inequality of Problem 7 may then be rewritten without the offending ellipsis as

$$\sum_{i \in I_n} \frac{1}{n+i} < \frac{3}{4}.$$

One often hears the phrase "the number of elements of such-and-such a set," where the number in question is a natural number. How exactly does one assign such a number to a set?

Consider by way of illustration the following "real-life" problem: What is the simplest method of determining whether there are more or fewer ladies than gentlemen in a ballroom? Of course, one might simply enumerate the ladies and gentlemen separately and compare the resulting numbers, but there is an easier and more natural method. If "ladies' choice" is announced, then each lady will rush to find herself a partner, and the question will then be decided by whether ladies or gentlemen remain standing along the wall.

Thus we shall say that a set A has k elements if there is a one-to-one correspondence (bijection) between A and the set I_k. We then need to show that a set cannot have (altogether) both three and four elements, for instance. Suppose that there are one-to-one correspondences $A \cong I_k$ and $A \cong I_l$, where $k < l$, say. It does not follow that just because I_k is a proper subset of I_l there is no one-to-one correspondence between them; after all there *is* a one-to-one correspondence between $\mathbb{N}$ and $\mathbb{N} \setminus \{1\}$. To this one may object that the sets I_k and I_l are finite; but then what is a finite set?$\ldots$

Theorem 8. *If $k \neq l$, then $I_k \ncong I_l$.*

Exercise. Prove this theorem using the principle of mathematical induction. □

Exercise. Find a one-to-one correspondence between the half-open interval $[0, 1)$ and the closed interval $[0, 1]$.

We next need the following lemma.

Lemma 3. *For all* $k, l \in \mathcal{N}$ *we have* $I_k \cong I_{k+l} \setminus I_l$.

To prove this, consider the map $\psi : I_k \to \mathcal{N}$, defined by $\psi(i) = i + l$. Since here $1 \leq i \leq k$, it follows that $l + 1 \leq \psi(i) \leq k + l$, so that for all $i \in I_k$ we have $\psi(i) \in I_{k+l} \setminus I_l$. It is not difficult to see that ψ is injective. To see that it is surjective, observe that if $j \in I_{k+l} \setminus I_l$, then $j > l$, whence $j = l + i$ for some i satisfying, as one sees easily, $i \leq k$. □

In what follows we shall denote by $|A|$ the number of elements of the set A (if there *is* such a natural number, i.e., if A is "finite").

Corollary. *If A and B are disjoint (finite) sets, then $|A \cup B| = |A| + |B|$. Hence in particular if B is a singleton, i.e., has just one element, then $|A \cup B| = s(|A|)$.*

Exercise. Set $H(|A|, |B|) := |A| \times |B|$. Prove that this formula defines a binary operation on the set $\mathcal{N}$ with properties M1 and M2, whence it follows that $|A \times B| = |A||B|$.

This corollary and exercise lend themselves naturally to the construction of the following archetypal but very naive "model" of the natural numbers: We shall take each natural number to be an appropriate collection(!), or pile, of sticks. Throwing one more stick on the pile gives us the next natural number, the successor. Addition and multiplication are then performed essentially as suggested by the above corollary and exercise. The reader is invited to try to formalize this approach.

The final problem in this chapter is specifically formulated so as to give the impression that induction, or rather inductive reasoning, is not appropriate for solving it.

Problem 8. Prove that if a, b, c, d, e are any real numbers satisfying $a \geq b \geq c \geq d \geq e$, then

$$a^2 - b^2 + c^2 - d^2 + e^2 \geq (a - b + c - d + e)^2.$$

(First prove the analogous inequality for three real numbers a, b, c satisfying $a \geq b \geq c$, by subtracting c^2 from both sides and factoring, and then apply this result twice to get the desired inequality.)

We conclude the chapter with a joke-theorem.[1] If a work of fiction should be true to life, then a good joke should surely contain an element of the truth. . . .

Theorem. *Alexander the Great never existed!*

We first establish an auxiliary result:

Lemma (false). *All objects are of the same color.*

The proof is by induction on the number n of objects being considered at any one time. The case $n = 1$ is trivial: clearly in a collection consisting of just one (monochromatic) object, there is only one color to be observed. If any k objects

---

[1] From "A Stress Analysis of a Strapless Evening Gown," Englewood Cliffs, N.J., 1963.

are assumed to have the same color, then surely it follows that any $k + 1$ objects must be of that color: for if we remove one object from a set of $k + 1$ objects, then the k remaining have the same color by assumption; but now if we put that one back and remove some other, then (again by the inductive hypothesis) it must have had the same color as the rest! □

Now to the proof of the theorem. Suppose that on the contrary Alexander the Great actually lived. According to historians (who never lie), Alexander rode a black horse called Bucephalus. Now, it is well known that there exist white objects,[2] so by the lemma all objects are white. Hence Alexander could not possibly have ridden on a black horse. This contradiction completes the proof. □

It is not difficult to train highschool students to give proofs by induction along the lines of the solution of Problem 1. However, as far as their mathematical education is concerned it is much more important (and infinitely more difficult) to inculcate in them the idea of inductive reasoning, which characterizes generally the search for solutions of mathematical problems (see, for instance, the solution of Problem 2), depending as this does essentially on an ability to divine from an examination of special cases the outline of an idea for a solution. Note that Problem 1 may be made more difficult by reformulating it as follows: Prove that the sum of the cubes of the first n natural numbers is a perfect square. Clearly, in this guise it is not suitable for the more junior students; however, for them there is an appropriate alternative: Show that the sum of the first n odd numbers is a perfect square. Incidentally, the identity obtained in solving the latter problem has an attractive geometrical interpretation. (What about Problem 1?) In the spirit of the "pigeonhole principle," one might advocate the "pine-tree principle" to younger students: In order to climb to the top of a pine tree, one has first somehow or other to get up onto the bottom branch, and then by some means climb from each branch to the next higher up.

The author has in this chapter included only a rather small selection of problems, for the most part chosen to illustrate the subject-matter, for the simple reason that in almost all of the later chapters there are assertions whose proofs use the method of mathematical induction. (See also the book [30], which contains a wealth of good problems of this type.) It is worth pointing out once more that very often the principle of mathematical induction (as the fundamental axiom for the natural numbers) is used without explicit mention, being indicated, for instance, merely by an ellipsis or by the words "and so on." In most such cases the teacher need not (or perhaps even *should* not) call the class's attention to the induction involved, but she herself *must* be aware of it.

A few words on Peano's axioms: Clearly, in teaching the axiomatic method to highschool students, familiar examples must be used, the foundations of geometry being the most common. However, the axiom-system for Euclidean geometry is rather complicated, to the extent that it is in practice impossible within the limits of the highschool curriculum to give even in outline a complete exposition of the elements of geometry—even Hilbert's definition of a ray is difficult for students to grasp (see [14]). Weyl's axiom-system is in this respect exceptional; however, there the subject of concern is more linear algebra than geometry. . . . Peano's axiom-system, on the other hand, structurally so simple and concise, concerned as it is with an object so familiar to everyone, and so closely bound up with the "method

---

[2]Mark Twain, "The Stolen White Elephant." In: *The Complete Short Stories of Mark Twain*, Doubleday, New York, 1957.

of mathematical induction" (in one form or another an item in highschool mathematical curricula), allows one to look afresh at the natural numbers and the operations on them. In view of this it seems somewhat odd that this axiom-system has not hitherto found a place in the mathematical programes of (specialized) schools. The author of this book was one of the few lucky ones: I was introduced to Peano's axioms as a senior highschool student, and although more than a quarter-century has gone by since then, I still recall the vivid impression they made on me.

CHAPTER 2

Combinatorics

2.1 Elementary problems

Combinatorics is the art of counting sets of configurations of things of one sort or another (or sometimes, in more complicated contexts, is concerned rather with the problem of the existence of certain configurations, or with finding the optimal one, in some sense, among all possible ones of a certain type).

This section contains combinatorial problems of an essentially elementary character, accessible to beginning highschool students [36]. The underlying purpose of these problems is to accustom the student to using the "sum rule" and "product rule" concerning respectively the union of two (disjoint) sets and the set of ordered pairs of elements of two sets, where the first member is from one of the sets and the second from the other, i.e., the "Cartesian product" (or "set product") of the given two sets in some order (see the corollary and exercise following Lemma 3 of Chapter 1).

Problem 1. How many three-digit numbers are divisible either by 9 or 15?

(Use the "inclusion–exclusion" formula $|A \cup B| = |A| + |B| - |A \cap B|$.)

Problem 2. By how many routes can one travel from town A to town B, passing through at most one other town on the way, if the towns are linked by roads as shown in the diagrams a and b?

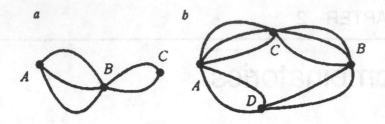

(The solutions of these two problems involve the sum and product rules in their purest form.)

Problem 3. In how many ways can 8 rooks be placed on a chessboard so that none threatens any other?

(In this rather easy problem we encounter the quantity $n! := 1 \cdot 2 \cdots \cdot n$ ("n-factorial"), important in what follows.)

Problem 4. Between Manchester and Liverpool there are two main highways linked at intervals by 10 secondary roads. In how many different ways can one travel from Manchester to Liverpool without passing through any junction twice?

One can leave Manchester by either highway, and then at each junction with a secondary road one can either take that road or stay on the same highway. Thus there are altogether 11 occasions where one has a choice between two possibilities, so that there are in total $2^{11} = 2048$ different routes from each city to the other.

An essentially similar solution involves the formulation of what we shall call an "isomorphic model" of the given problem. Consider all strings (finite sequences) of eleven 0s and 1s (i.e., 11-position binary numbers); there are 2^{11} such strings. With each of these strings we associate a route from Manchester to Liverpool as follows: If the string begins with a 1, then the corresponding route leaves Manchester by the northern highway (see the diagram), and for each $i = 2, 3, \ldots, 11$, if the ith position is occupied by a 1, then the route goes by the $i - 1$st secondary road. Thus, for example the route shown in the diagram corresponds to the string of eleven 1s. Clearly, this correspondence is one-to-one, whence the number of routes is equal to the number of 11-position binary strings.

Problem 5. In how many ways can the white king and black king be placed on a chessboard (in accordance with the rules of chess)?

Answer. In $4 \cdot (64 - 4) + 24 \cdot (64 - 6) + 36 \cdot (64 - 9) = 3612$ ways: If the black king is on one of the four corner squares, then there are $64 - 4 = 60$ squares where the white king can go; if the black king is on a border square not at a corner (there are 24 such squares), then there are $64 - 6$ squares available to the white king;

finally, for each of the 36 interior squares there are $64 - 9 = 55$ squares where the white king can be placed.

The next problem has an unexpected answer (at first glance).

Problem 6. Among the 8-digit numbers are there more with or without 1 as a digit?

There are altogether $9 \cdot 10^7$ eight-digit numbers. (Note that the leftmost digit cannot be zero.) Those where 1 does not occur as a digit number in all $8 \cdot 9^7$. Since

$$\frac{9 \cdot 10^7}{8 \cdot 9^7} = \frac{9}{8}(1 + \frac{1}{9})^7 > \frac{9}{8}(1 + \frac{7}{9}) = 2,$$

(where we have used Bernoulli's inequality—see Chapter 5), the 8-digit numbers without 1 as a digit account for fewer than half of all 8-digit numbers.

Problem 7. How many different divisors (including 1 and the number itself) does each of the following numbers have: (a) 720; (b) $n = p_1^{s_1} p_2^{s_2} \ldots p_k^{s_k}$ (where the p_l are distinct primes)?

This is typical of those problems where in the formulation of the general case there is a hint; thus here (b) is easier than (a).

Problems like those given above are legion, and moreover may be multiplied indefinitely by means of different, though isomorphic, formulations.

Problem 8. A class consists of 9 girls and 16 boys. In how many ways can a pair of students be chosen to compete in a tennis tournament of: (a) mixed doubles; (b) women's doubles; (c) men's doubles?

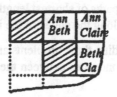

The product rule gives the answer to (a). To answer (b), one may either draw up a table (as in the diagram), or else argue as follows: If we imagine sticking labels marked "1" and "2" on each pair of girls, then the number of *labeled* such pairs (i.e., ordered pairs of girls) is clearly $9 \cdot 8 = 72$, and then since each *unordered* pair (doubleton) gives rise to two ordered pairs, the number of possible selections of a team to compete in a women's doubles tournament is half of 72, or 36.

Here are two further questions using these ideas.

Problem 9. How many anagrams (i.e., rearrangements of the letters) are there of each of the following words: (a) barb; (b) radar; (c) baobab.

Problem 10. In how many ways can two rubies, two emeralds, and two garnets be arranged: (a) in a row; (b) in a circle; (c) to form a necklace? (How does an arrangement of these stones in a circle differ from their arrangement as a necklace?)

In the last two problems we might have introduced terminology such as "arrangements with repetitions," etc. However, in the author's opinion if one is encountering a concept for the first time, then an excess of terminology tends to impede understanding.

Problem 11. (a) How many different dice are there? (b) What if the dice are tetrahedral in shape rather than cubical? (We have in mind a tetrahedron with one, two, three, and four spots on its four faces.)

Here is a solution of (b). Set such a tetrahedral die down on the face with just one spot, and then turn it so that the face with two spots is towards you. This fixes the position of the tetrahedron, and one now sees that there are two distinct ways of situating the 3-spot and 4-spot on the remaining two faces. Hence there are just two different tetrahedral dice, a "right-handed" and a "left-handed" one.

2.2 Combinations and recurrence relations

We begin with three problems.

Problem 12. In how many ways can three (distinct) things be chosen from among seven?

Problem 13. How many 7-term sequences are there containing three 0s and four 1s?

Problem 14. Consider the Euclidean plane furnished with a rectangular Cartesian coordinate system. How many paths of shortest length are there from the origin O to the point $(4, 3)$ proceeding along the edges of the integral lattice?

Exercise. Show that the preceding three problems are isomorphic, i.e., that there are natural one-to-one correspondences between the sets of configurations to be counted in the three situations.

For some, the solution of the first of these problems may seem easiest: The number of arrangements (in a row) of three things from a set of seven things is $7 \cdot 6 \cdot 5 = 210$, since there are 7 ways of choosing the first thing, and for each such choice there are 6 choices for second place, and corresponding to each choice for the first two places there are 5 ways of filling the third place. Now, each subset of three things can be ordered (permuted) in 3! ways and so contributes this number to the above total of 210; hence the number of *unordered* 3-element subsets of a 7-element set is $\frac{7 \cdot 6 \cdot 5}{3!}$. In general, the number of k-element subsets of an n-element set is $\frac{n \cdot (n-1) \cdot \ldots \cdot (n-k+1)}{k!} = \frac{n!}{k!(n-k)!}$, and this important number is denoted by $\binom{n}{k}$ ("n-choose-k").

On the other hand, for other tastes, the geometrical flavor of the last of the above three problems may make its solution easier to swallow: Consider the following three points: $A(k, n-k)$, $B(k-1, n-k+1)$, $C(k, n-k+1)$. Clearly, the number of shortest paths from O to C is equal to the number of such paths from O to A plus the number from O to B (see Figure a below). This makes it easy to progressively label the points of the integral lattice by moving outwards from O, each point being labeled with the number of shortest paths from O to that point (as in Figure b below). The numbers so obtained are usually arranged somewhat differently—to form "Pascal's triangle" (Figure c). If we denote the $(k+1)$st number from the left in the $(n+1)$st row of Pascal's triangle by C_n^k ($n, k = 0, 1, \ldots$), then the above relationship between numbers of paths becomes $C_{n+1}^k = C_n^k + C_n^{k-1}$: each entry in Pascal's triangle is the sum of the two entries immediately above it.

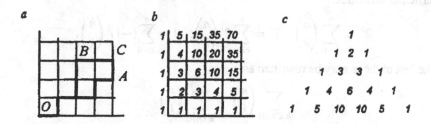

Exercise. Verify that the number $\binom{n}{k}$ satisfies the above recurrence relation, both directly using the formula for $\binom{n}{k}$ given above, and via its interpretation as the number of "combinations" of k objects taken from a set of n.

By fiat one sets $0! := 1$. Then since $\binom{n}{0} = C_n^0 = \binom{n}{n} = C_n^n$, it follows that $\binom{n}{k} = C_n^k = \frac{n!}{k!(n-k)!}$. (How?)

Problem 15. Is it possible for every resident of China to have a distinct set of teeth (i.e., one toothless person, at most 32 with exactly one tooth, etc.)?

There are two approaches to this problem. The number of ways in which one can have exactly k teeth remaining in one's jaw is $\binom{32}{k}$, so that the total number of possible sets of teeth is given by $\sum_{k=0}^{32} \binom{32}{k}$. Alternatively, observe that for each tooth there are just two possibilities: either it is still there, or there is a gap where it used to be. Hence there are altogether $2^{32} = 4,294,864,896$ possible sets of teeth, which number exceeds by far the population of China.

As a result of solving this problem in two ways, we have incidentally established the equality $\sum_{k=0}^{32} \binom{32}{k} = 2^{32}$.

Exercise. Prove that the total number of subsets of a set of n elements is 2^n. Find another expression for this number by counting the k-element subsets for each $k = 0, 1, \ldots, n$.

As a result of doing this exercise, one obtains the general version of the above equality, namely $\sum_{k=0}^{n} \binom{n}{k} = 2^n$. Checking a few cases lends plausibility to the similar identity $\binom{n}{0} + 2\binom{n}{1} + 4\binom{n}{2} + \cdots + 2^n\binom{n}{n} = 3^n$. The reader familiar with the binomial theorem may by now have realized that we are about to consider the numbers $\binom{n}{k}$ in their role as binomial coefficients.

Problem 16. Denote by D_n^k the coefficient of the monomial $a^{n-k}b^k$ in the expansion of $(a + b)^n$. Using $(a + b)^n = (a + b)(a + b)^{n-1}$, prove that $D_{n+1}^k = D_n^k + D_n^{k-1}$.

From this and the fact that $D_n^0 = D_n^n = 1$, it follows that $D_n^k = \binom{n}{k}$.

Exercise. Establish directly an isomorphism between Problem 16 on the binomial coefficients and Problems 12–14 (generalized).

From the binomial theorem just proved (by the reader!), we obtain the following particular identities:

$$2^n = \sum_{k=0}^n \binom{n}{k}; \quad 3^n = \sum_{k=0}^n 2^k \binom{n}{k}; \quad 0 = \sum_{k=0}^n (-1)^k \binom{n}{k}.$$

The last of these may be rewritten as

$$\sum_{k \text{ even}} \binom{n}{k} = \sum_{k \text{ odd}} \binom{n}{k}.$$

There are a great many other relations among the binomial coefficients, many of which can be obtained by exploiting properties of the function $P_n(x) :=$ $\sum_{k=0}^n \binom{n}{k} x^k$.

Lemma 1. *The identity $P_{n+1}(x) = (x + 1)P_n(x)$ is equivalent to $\binom{n+1}{k} = \binom{n}{k} + \binom{n}{k-1}$.*

This is immediate from $(x + 1) \sum_{k=0}^n \binom{n}{k} x^k = \sum_{k=0}^{n+1} \left(\binom{n}{k} + \binom{n}{k-1} \right) x^k$. $\square$

Corollary. *We have $P_n(x) = (x + 1)^n$.*

(This also follows more directly from the definition of $P_n(x)$ and the binomial theorem. Alternatively, one may consider this as affording an alternative proof of the binomial theorem.)

Here the word "function" was not used lightly. Since

$$\frac{dP_n(x)}{dx} = \sum_{k=1}^n \binom{n}{k} kx^{k-1} = \sum_{k=0}^{n-1} \binom{n}{k+1}(k + 1)x^k,$$

while on the other hand

$$\frac{dP_n(x)}{dx} = n(x + 1)^{n-1} = n \sum_{k=0}^{n-1} \binom{n-1}{k} x^k,$$

we see that

$$\binom{n}{k+1} = \frac{n}{k+1} \binom{n-1}{k} = \frac{n(n-1)}{(k+1)k} \binom{n-2}{k-1} = \cdots$$

$$= \frac{n(n-1)\cdots(n-k+1)}{(k+1)k \cdots 2} \binom{n-k}{1}$$

$$= \frac{n(n-1)\cdots(n-k)}{(k+1)!} = \frac{n!}{(n-k-1)!(k+1)!},$$

which may be considered as affording yet another (relatively longwinded!) verification of the Binomial Theorem. Of main interest here is the method of proof.

Exercise. Prove that $\binom{2n}{k} = \sum_{l=0}^{k} \binom{n}{l}\binom{n}{k-l}$ using $((1+x)^n)^2 = (1+x)^{2n}$.

Problem 17. Prove that $\tan 3x = \frac{3\tan x - \tan^3 x}{1 - 3\tan^2 x}$. Find and verify the analogous formula for $\tan nx$.

Set $t := \tan x$. Since $\tan 2x = 2t/(1 - t^2)$, $\tan 3x = (3t - t^3)/(1 - 3t^2)$, and $\tan 4x = (4t - 4t^3)/(1 - 6t^2 + t^4)$ (verify!), one is led (eventually) to the general formula

$$\tan nx = \frac{\sum_{k=0}^{[n/2]}(-1)^k \binom{n}{2k+1}t^{2k+1}}{\sum_{k=0}^{[n/2]}(-1)^k \binom{n}{2k}t^{2k}},$$

which is not difficult to prove by means of a direct induction. Use of De Moivre's theorem, however, allows a slicker proof. Since

$$\cos nx + i \sin nx = (\cos x + i \sin x)^n = \sum_{k=0}^{n} \binom{n}{k} i^k \sin^k x \, \cos^{n-k} x,$$

it follows that

$$\cos nx = \sum_{k=0}^{[n/2]}(-1)^k \binom{n}{2k} \sin^{2k} x \, \cos^{n-2k} x,$$

and

$$\sin nx = \sum_{k=0}^{[n/2]}(-1)^k \binom{n}{2k+1} \sin^{2k+1} x \, \cos^{n-2k-1} x,$$

whence

$$\tan nx = \frac{\sin nx}{\cos nx} = \frac{\sum_{k=0}^{[n/2]}(-1)^k \binom{n}{2k+1} \sin^{2k+1} x \, \cos^{n-2k-1} x}{\sum_{k=0}^{[n/2]}(-1)^k \binom{n}{2k} \sin^{2k} x \, \cos^{n-2k} x}$$

$$= \frac{\sum_{k=0}^{[n/2]}(-1)^k \binom{n}{2k+1}t^{2k+1}}{\sum_{k=0}^{[n/2]}(-1)^k \binom{n}{2k}t^{2k}}.$$

Elementary combinatorial problems may often be highly nontrivial to solve. An instance of this is afforded by a problem concerning the enumeration of all graphs of a certain type, which we shall now describe. A graph is called *labeled* if with each of its vertices there is associated a number from the set $I_n := \{1, 2, \ldots, n\}$ in such a way that distinct vertices are labeled with different numbers and every number in I_n is used (i.e., there is given a bijection from I_n to the set of vertices).

Theorem 1 ((Cayley) [37]). *There are exactly n^{n-2} distinct (i.e., nonisomorphic) labeled trees with n vertices.*

Exercise. Calculate the number of different labelings of each of the following two (unlabeled) trees:

Answer. Respectively 12 and 4, making a total of $16 = 4^2$.

To prove the theorem, fix on some number s from I_n, and for each $k = 1, \ldots, n-1$, denote by $S(n, k)$ the number of labeled trees with n vertices in which the vertex labeled with s has degree (or valency) $d(s) = k$, i.e., has k edges incident with it. Given any labeled tree A with n vertices and with the vertex u, say, labeled by s of degree $k - 1$, we construct a labeled tree B as follows: Choose any two adjacent vertices v, w of A other than u (assuming $n \geq 4$) and remove the edge joining v and w. This results in two trees, one of which contains the vertex u (labeled s) and just one of the vertices v, w, say v, and the other the vertex w. Now join u and w by means of a new edge to obtain a new labeled tree B where the vertex u now has degree k. We shall call the pair (A, B) *cognate*. Since there are $S(n, k-1)$ labeled trees of the form A, and corresponding to each of these there are $(n - 1) - (k - 1)$ choices for the edge to be removed in constructing an associated labeled tree B, the cognate pairs (A, B) number altogether $(n - k)S(n, k - 1)$. We now recalculate the number of such pairs using a different method. To this end consider any labeled tree B with n vertices in which the vertex labeled s has degree k; denote this vertex by v, and the vertices adjacent to v by $w_1, \ldots, w_k$. We now perform operations on B yielding a cognate labeled tree A: first remove an edge incident with v, say the edge joining v to w_i (this can be done in k ways), and then re-attach the resulting two subtrees by joining w_i by an edge to any vertex of the subtree containing v (feasible in $n - 1 - n_i$ ways, where n_i denotes the number of vertices of the subtree containing w_i). This construction yields a cognate labeled tree A, and every such tree can be obtained in this manner. Hence their are in all $(n - 1 - n_1) + \cdots + (n - 1 - n_k) = (n - 1)(k - 1)$ cognates A of each B, whence the total number of cognate pairs (A, B) is $(n - 1)(n - k)S(n, k)$. We have thus established the equation $(n - 1)(n - k)S(n, k) = (n - k)S(n, k - 1)$.

Exercise. Deduce that $S(n, k) = \binom{n-2}{k-1}(n - 1)^{n-k-1}$.

(Use the fact that $S(n, n - 1) = 1$.)
Hence the number of labeled trees on n vertices is given by

$$\sum_{k=1}^{n-1} S(n, k) = \sum_{k=1}^{n-1} \binom{n - 2}{k - 1}(n - 1)^{n-k-1} = ((n - 1) + 1)^{n-2} = n^{n-2}.$$

□

We conclude this section with two simple but significant problems.

Problem 18. Prove that the product of any k consecutive natural numbers is divisible by $k!$.

It is easy to see that such a product is divisible by every natural number $\leq k$. However, of course a number with the latter property need not be divisible by $k!$.

Problem 19. Prove the inequality $\binom{n}{k} \leq \binom{n}{[n/2]}$, $k = 1, \ldots, n$.

2.3   Recurrence relations and power series

In the preceding section we glimpsed the possible usefulness of associating a function (there it was a polynomial, but more generally power series turn out to be appropriate) with a sequence of numbers satisfying a recurrence relation. We shall now illustrate this technique as it applies to the Fibonacci sequence, traditionally arrived at as follows:

Each pair of rabbits produces each month a pair of offspring, which in turn commences reproducing starting from the second month of its existence. How many pairs of rabbits will there be after n months, originating from a single newborn pair, if we neglect deaths?

If we denote by a_n the number of pairs of rabbits after n months, then clearly $a_0 = a_1 = 1$ (as no issue in the first month), and for $n \geq 2$, $a_n = a_{n-1} + a_{n-2}$ (since in addition to the pairs existing in the preceding month we now have the issue of those born at least a month prior to that).

Theorem 2. *The formula for* a_n *is as follows:*

$$a_n = \frac{1}{\sqrt{5}}\left(\left(\frac{1+\sqrt{5}}{2}\right)^{n+1} - \left(\frac{1-\sqrt{5}}{2}\right)^{n+1}\right).$$

To prove this we first form the "generating function" of the Fibonacci sequence $\{a_n\}_{n=0}^{\infty}$, i.e., the power series $\phi(x) = \sum_{n=0}^{\infty} a_n x^n$, and then calculate the product

$$(1 - x - x^2)\phi(x) = (1 - x - x^2)\sum_{n=0}^{\infty} a_n x^n$$

$$= (1 - x - x^2)(a_0 + a_1 x + a_2 x^2 + \cdots)$$

$$= a_0 + (a_1 - a_0)x + (a_2 - a_1 - a_0)x^2 + \cdots.$$

In view of the initial conditions and the recurrence relation for the Fibonacci numbers, the last series here is actually just 1, whence $\phi(x) = 1/(1 - x - x^2)$. Since $1 - x - x^2 = -(\alpha_1 - x)(\alpha_2 - x)$, where $\alpha_1 = -\frac{(1-\sqrt{5})}{2}$, $\alpha_2 = -\frac{(1+\sqrt{5})}{2}$, and

$$\frac{1}{\alpha - x} = \frac{1/\alpha}{1 - x/\alpha} = \frac{1}{\alpha}\sum_{k=0}^{\infty}\left(\frac{x}{\alpha}\right)^k = \sum_{k=0}^{\infty}\frac{x^k}{\alpha^{k+1}},$$

it follows that

$$\sum_{k=0}^{\infty} a_k x^k = \frac{1}{1 - x - x^2} = -\frac{1}{(\alpha_1 - x)(\alpha_2 - x)}$$

$$= \frac{1}{(\alpha_1 - \alpha_2)} \left(\frac{1}{\alpha_1 - x} - \frac{1}{\alpha_2 - x} \right) = \frac{1}{(\alpha_1 - \alpha_2)} \sum_{k=0}^{\infty} \left(\frac{x^k}{\alpha_1^{k+1}} - \frac{x^k}{\alpha_2^{k+1}} \right)$$

$$= \frac{1}{\sqrt{5}} \sum_{k=0}^{\infty} ((-\alpha_2)^{k+1} - (-\alpha_1)^{k+1}) x^k$$

$$= \sum_{k=0}^{\infty} \frac{1}{\sqrt{5}} \left(\left(\frac{1 + \sqrt{5}}{2} \right)^{k+1} - \left(\frac{1 - \sqrt{5}}{2} \right)^{k+1} \right) x^k.$$

Comparison of coefficients of like powers of x now yields the desired formula. $\square$

This argument is undoubtedly elegant, although it has the air of a conjuring trick. A little explanation might seem to be called for at one or two places. For example, what meaning should be given to the equation $\phi(x) = 1/(1 - x - x^2)$? This seems to make no sense for $x = 1$, for instance.

To clarify this we consider the *ring of formal power series*

$$\mathbb{R}[[x]] := \left\{ \sum_{k=0}^{\infty} a_k x^k \mid a_k \in \mathbb{R} \right\},$$

with addition and multiplication defined by:

$$\sum_{k=0}^{\infty} a_k x^k + \sum_{k=0}^{\infty} b_k x^k := \sum_{k=0}^{\infty} (a_k + b_k) x^k;$$

$$\sum_{k=0}^{\infty} a_k x^k \cdot \sum_{k=0}^{\infty} b_k x^k := \sum_{k=0}^{\infty} c_k x^k, \quad \text{where } c_k := \sum_{i=0}^{k} a_i b_{k-i}.$$

Exercise. Prove that the set $\mathbb{R}[[x]]$ with these two operations is a commutative ring. Show also that the units (i.e., multiplicatively invertible elements) of this ring are precisely those series $\sum_{k=0}^{\infty} a_k x^k$ with $a_0 \neq 0$.

Note that the multiplicative identity element of this ring is the series $1 + 0 + \cdots$, which we shall identify with the number $1 \in \mathbb{R}$. The problematical identity $\phi(x)(1 - x - x^2) = 1$ of the above proof may now be interpreted as meaning simply that $\phi(x) = (1 - x - x^2)^{-1}$ in the ring of formal power series. Similarly, for $i = 1, 2$, $\phi_i(x) := \sum_{k=0}^{\infty} x^k / \alpha_i^{k+1} = (\alpha_i - x)^{-1}$ in that ring.

Exercise. Prove that in any commutative ring if u, v, and $u - v$ are all units, then $(uv)^{-1} = (u - v)^{-1}(v^{-1} - u^{-1})$.

We are now in a position to make sense of the calculation carried out in the proof of Theorem 2. Taking $1/u(x)$ to mean the inverse of $u(x)$ in the ring $\mathbb{R}[[x]]$, we have

$$\phi(x) = (1 - x - x^2)^{-1} = -((\alpha_1 - x)(\alpha_2 - x))^{-1}$$
$$= (\alpha_2 - \alpha_1)^{-1}((\alpha_2 - x)^{-1} - (\alpha_1 - x)^{-1})$$
$$= \frac{1}{\alpha_1 - \alpha_2}(\phi_1(x) - \phi_2(x)).$$

Here is another interesting method of arriving at the explicit formula for the Fibonacci numbers [2]. Setting

$$\xi_n := \begin{pmatrix} a_{n-1} \\ a_n \end{pmatrix} \in \mathbb{R}^2,$$

we have

$$\xi_n = \begin{pmatrix} a_{n-1} \\ a_n \end{pmatrix} = \begin{pmatrix} a_{n-1} \\ a_{n-1} + a_{n-2} \end{pmatrix} = \begin{pmatrix} 0 & 1 \\ 1 & 1 \end{pmatrix} \begin{pmatrix} a_{n-2} \\ a_{n-1} \end{pmatrix}$$
$$= A\xi_{n-1} = A^{n-1}\xi_1 = A^{n-1} \begin{pmatrix} 1 \\ 1 \end{pmatrix} = A^n \begin{pmatrix} 0 \\ 1 \end{pmatrix},$$

whence we see that it suffices to find the explicit form of the nth power of the matrix A. Now, the characteristic polynomial of that matrix is

$$\det(A - tI) = \begin{vmatrix} -t & 1 \\ 1 & 1 - t \end{vmatrix} = t^2 - t - 1,$$

whence its eigenvalues $\lambda_{1,2} = (1 \pm \sqrt{5})/2$. Since the matrix A has distinct eigenvalues, it (or rather the linear transformation it represents) will have diagonal form relative to a suitable basis for $\mathbb{R}^2$, so that there is a matrix P such that $A = P^{-1}DP$, where $D = \operatorname{diag}(\lambda_1, \lambda_2)$.

Exercise. Prove that $A^n = P^{-1}D^nP$.

Hence a_n has the form $a_n = c_1\lambda_1^n + c_2\lambda_2^n$. The coefficients c_1 and c_2 may now be inferred from the initial conditions $a_0 = a_1 = 1$, since these yield $c_1 + c_2 = 1$ and $c_1\lambda_1 + c_2\lambda_2 = 1$.

2.4 Generating functions

We begin with the following problem.

Problem 20. Prove that a 3×3 magic square must have the number 5 in the central position, and 2, 4, 6, and 8 in the corners.

Since the sum of all entries in a 3×3 magic square is $1 + 2 + \cdots + 9 = 45$, each row, column, and diagonal must add up to 15. Listing all possible partitions of 15 into a sum of three different numbers between 1 and 9, we obtain: $1 + 5 + 9, 1 + 6 + 8, 2 + 4 + 9, 2 + 5 + 8, 2 + 6 + 7, 3 + 4 + 8, 3 + 5 + 7, 4 + 5 + 6$.

The number at the center of the magic square has to occur in four such partitions (see the diagram), so that 5 is the only candidate. Each corner entry participates in just three partitions, and so again by inspection of the list we see that these must be 2, 4, 6, 8.

Problem 21. Find the number of *ordered* partitions of a natural number n as a sum of k nonnegative integers.

Answer. $\binom{n+k-1}{k-1}$.

The main goal of the present section is the substantial one of finding relations among the numbers of partitions of natural numbers (into sums of natural numbers), using the method of generating functions [26]. The classical problem of calculating these numbers is of considerable difficulty. For instance, the number of partitions of 100 is 190569292.

We denote by $p(n)$ the number of partitions of the natural number n, and extend the domain of p to 0 by setting $p(0) := 1$.

Lemma 2. *The following factorization is valid:*

$$\sum_{n=0}^{\infty} p(n)x^n = (1 + x + x^2 + \cdots)(1 + x^2 + x^4 + \cdots)\cdots$$

$$\times (1 + x^k + x^{2k} + \cdots)\cdots.$$

(Note that although the product on the right-hand side of this equation has infinitely many factors, calculation of the coefficient of any particular power of x requires only a finite number of operations to be performed.)

To prove the lemma, one associates with each finite sequence $(m_1, m_2, \ldots, m_k)$ of nonnegative integers on the one hand the (formal) product $x^{m_1} x^{m_2} \cdots x^{m_k}$, and on the other hand the partition of the number $n = m_1 + 2m_2 + \cdots + km_k$ into m_1 ones, m_2 twos, etc. Since clearly both of these correspondences are one-to-one, the coefficient of x^n obtained by expanding the product on the right-hand side of the equation in the lemma will indeed be equal to the number $p(n)$ of partitions of n. □

Corollary. *The following identity is valid:*

$$\sum_{n=0}^{\infty} p(n)x^n = \frac{1}{(1-x)(1-x^2)\cdots}.$$

We consider now two special sorts of partition: Denote by $l(n)$ the number of partitions of n into odd terms only, and by $d(n)$ the number of partitions where the terms are all distinct.

Exercise. Prove that

$$\sum_{n=0}^{\infty} l(n)x^n = \frac{1}{(1-x)(1-x^3)\cdots}, \qquad \sum_{n=0}^{\infty} d(n)x^n = (1+x)(1+x^2)\cdots.$$

Theorem 3. *The numbers of partitions of n of each of the above special types are equal:* $d(n) = l(n)$.

Denoting by $D(x)$ and $L(x)$ the generating functions for the sequences of numbers $d(n)$ and $l(n)$ respectively, we have

$$D(x) = \sum_{n=0}^{\infty} d(n)x^n = (1+x)(1+x^2)\cdots$$

$$= \frac{(1-x^2)}{(1-x)}\frac{(1-x^4)}{(1-x^2)}\frac{(1-x^6)}{(1-x^3)}\cdots = \frac{1}{(1-x)(1-x^3)\cdots} = L(x). \quad \square$$

Notice, however, that the last line of this proof involves infinitely many cancellations in a product of infinitely many fractional expressions. We shall therefore now give a more rigorous, though less elegant, argument:

Two elements $f, g \in \mathbb{R}[[x]]$ are said to be *congruent modulo* x^n, written $f \equiv g \pmod{x^n}$, if $f - g = x^n h$ for some element $h \in \mathbb{R}[[x]]$. Clearly, $f = g$ precisely if $f \equiv g \pmod{x^n}$ for arbitrarily large n. Now

$$D(x) \equiv (1+x)\ldots(1+x^{2n}) \pmod{x^{2n+1}}$$

$$= (1-x^2)\cdots(1-x^{4n})(1-x)^{-1}\cdots(1-x^{2n})^{-1} \pmod{x^{2n+1}}$$

$$\equiv (1-x)^{-1}\cdots(1-x^{2n-1})^{-1} \pmod{x^{2n+1}} \equiv L(x) \pmod{x^{2n+1}}.$$

Hence $D(x) = L(x)$. $\quad\square$

(To put it in slightly more abstract algebraic terms, we have shown here that the natural images of $D(x)$ and $L(x)$ in each quotient ring $\mathbb{R}[[x]]/x^{2n+1}\mathbb{R}[[x]]$ coincide, from which it follows that these elements are themselves actually equal.)

Another approach to proving the completely nonobvious identity of Theorem 3 uses a property of so-called "Young tableaux." Let $n = (2k_1+1)+(2k_2+1)+\cdots+(2k_s+1)$, where $k_1 \geq k_2 \geq \cdots \geq k_s$, be any partition of n as a sum of odd (positive) numbers only. Consider a right angle along each of whose arms $k_1 + 1$ equally spaced points are marked starting at the apex, so that altogether there are $2k_1 + 1$ points marked. (In the diagram, where $k_1 = 4$, one arm extends horizontally to the right and the other vertically downwards.) Imagine now another right angle one space lower and one space to the right, with $2k_2+1$ similarly spaced points marked on it, and so on. (The diagram shows the arrangement of points corresponding to the partition (9; 7; 7; 5; 3) of 31.)

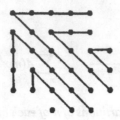

Next one draws broken lines each made up of two straight segments as illustrated in the diagram. There we see that these broken lines pass through 9, 6, 3, 8, 4, and 1 points respectively, thus yielding a partition of 31 into pairwise distinct numbers.

Exercise. Prove that the indicated algorithm always yields a partition of n into distinct summands. (It is more difficult to show that the resulting correspondence is bijective.)

Returning to the result established earlier to the effect that the generating function for the number of partitions is the inverse of the infinite product $(1 - x)(1 - x^2)(1 - x^3)\cdots$, we now do a little calculation: First we have $(1 - x)(1 - x^2) = 1 - x - x^2 + x^3$, and multiplying this by $1 - x^3$ yields $1 - x - x^2 + x^4 + x^5 - x^6$. Multiplication by further factors yields in turn

$$1 - x - x^2 + 2x^5 - x^8 - x^9 - x^{10},\quad 1 - x - x^2 + x^5 + \cdots.$$

Observe that the first four terms given explicitly in the last of these polynomials remain unaffected by further multiplications by the binomials $1 - x^k$, $k \geq 6$. With some perseverance one may calculate that the first eleven terms of the expanded infinite product are

$$1 - x - x^2 + x^5 + x^7 - x^{12} - x^{15} + x^{22} + x^{26} - x^{35} - x^{40} + \cdots.$$

Theorem 4 (Euler). *One has*

$$\prod_{k=1}^{\infty}(1 - x^k) = \sum_{q=-\infty}^{\infty}(-1)^q x^{(3q^2+q)/2}.$$

Corollary. *The following recurrence relation is valid:*

$$p(n) = p(n - 1) + p(n - 2) - p(n - 5) - p(n - 7) + p(n - 12) + \cdots.$$

This can be seen as follows: By the theorem, the fact established earlier that $P(x)\prod_{k=1}^{\infty}(1 - x^k) = 1$, can be rewritten as

$$\sum_{n=0}^{\infty} p(n)x^n \cdot (1 - x - x^2 + x^5 + x^7 - \cdots) = 1.$$

This yields for each $n \geq 1$,

$$x^n(p(n) - p(n - 1) - p(n - 2) + \cdots) = 0,$$

whence the desired formula. □

We now prove Theorem 4. Denote by a_n the coefficient of x^n in the expansion of the product $\prod_{k=1}^{\infty}(1-x^k)$. Clearly, a_n is the sum of the numbers $(-1)^k$ over all partitions of n as a sum of k distinct natural numbers. Consider an arbitrary such partition $n = n_1 + n_2 + \cdots + n_k$, with $n_1 < n_2 < \cdots < n_k$. We shall distinguish three different types of such partitions. To this end we associate with each such partition the largest integer s such that the last s members $n_{k-s+1}, n_{k-s+2}, \ldots, n_k$ of the partition are consecutive. Clearly, $1 \leq s \leq k$.

We define a Type I partition to be one for which $n_1 \leq s$, excluding those partitions for which $n_1 = s = k$, i.e., excluding partitions of the form $n = k + (k+1) + \cdots + (2k-1)$. Type II partitions are to be those for which $n_1 > s$, excluding those with $n_1 - 1 = s = k$, i.e., excluding partitions of the form $n = (k+1) + (k+2) + \cdots + 2k$. The excluded partitions then make up Type III. Observe that the last type exists only for n of the form $k + (k+1) + \cdots + (2k-1) = k(3k-1)/2$, or, in the other case, $k(3k+1)/2$, i.e., precisely those numbers appearing as exponents in the right-hand side of Euler's identity!

With each partition $n = n_1 + n_2 + \cdots + n_k$ of Type I we now associate the following partition (of the same number n):

$$n = n_2 + n_3 + \cdots + (n_{k-n_1+1} + 1) + \cdots + (n_k + 1).$$

This is feasible, since $n_1 < k$ for this type of partition. Since also here $n_1 \leq s$, the new partition will have $s' = n_1$ as the length of its largest terminal segment of consecutive numbers. Its first member is n_2, which exceeds n_1 and therefore s'. Hence the new partition is of Type II. This correspondence may be expressed in terms of Young tableaux, as shown in the diagram: the points of the top row are redistributed one each to the bottommost rows.

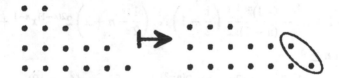

Thus to each k-partition of n of Type I there corresponds in this way a $(k-1)$-partition of Type II, and it is not difficult to verify that this correspondence is bijective. Hence in the process of expanding the infinite product, each contribution $(-1)^k$ to the coefficient a_n of x^n from a k-partition of Type I cancels with the contribution $(-1)^{k-1}$ from the corresponding $(k-1)$-partition of Type II. Consequently, the only powers x^n with nonzero coefficient are those where n has a partition of Type III. $\square$

Problem 22. Prove that in order to subdivide a convex n-gon into triangles by means of nonintersecting diagonals (i.e., not intersecting in the interior of the polygon), exactly $n-3$ such diagonals are required.

As our final application of the method of generating functions, we consider a problem from combinatorial geometry [26].

For each $n \geq 3$ denote by D_n the number of distinct subdivisions of an n-gon $A_1 A_2 \cdots A_n$ into triangles by means of nonintersecting diagonals. Set $D_2 := 1$.

Problem 23. Prove that for every $n \geq 3$ one has

$$D_n = D_2 D_{n-1} + D_3 D_{n-2} + \cdots + D_{n-1} D_2.$$

(The side $A_1 A_2$ of the given polygon is a side of some triangle of the subdivision, whose third vertex is one of the points $A_3, \ldots, A_n$; see the diagram.)

Consider the generating function $D(x) := D_2 x^2 + D_3 x^3 + \cdots$ of the sequence $\{D_n\}_{n=2}^{\infty}$.

Exercise. Show how it follows from the recurrence relation of Problem 23 that the generating function $D(x)$ of the sequence D_n satisfies the functional equation $y^2 - xy + x^3 = 0$.

Hence $y = x(1 \pm \sqrt{1 - 4x})/2$, and since the power series $D(x)$ has no degree-one term, we must have in fact $y = x(1 - \sqrt{1 - 4x})/2$, whence

$$\sum_{n=2}^{\infty} D_n x^n$$

$$= \frac{x}{2} \left(1 - 1 - \ldots - \frac{(-1)^{n-1}}{(n-1)!} \frac{1}{2} \left(\frac{1}{2} - 1 \right) \cdots \left(\frac{1}{2} - n + 2 \right) 2^{2(n-1)} x^{n-1} + \ldots \right),$$

yielding for $n \geq 3$

$$D_n = \frac{1 \cdot 3 \cdot \cdots \cdot (2n - 5) \cdot 2^{(n-2)}}{(n-1)!} = \frac{2 \cdot 6 \cdot \cdots \cdot (4n - 10)}{2 \cdot 3 \cdot \cdots \cdot (n - 1)}.$$

Exercise. Try to make sense of the above calculation.

2.5   The numbers π, e, and n-factorial

Exercise. Write $a_n := n! n^{-n}$. Prove that

$$\lim_{n \to +\infty} \frac{a_n}{a_{n+1}} = e.$$

As we have seen, the expression $n!$ often crops up in the answers to counting problems. The above exercise suggests that a more amenable function asymptotic to $n!$ should have as factors n^n and e^{-n}. The precise result is as follows [29]:

Theorem 5 (Stirling). *The following asymptotic equation holds:*

$$n! = \sqrt{2\pi n}\, n^n e^{-n}(1 + o(1)), \quad i.e., \quad \lim_{n \to +\infty} \frac{n! e^n}{\sqrt{2\pi n}\, n^n} = 1.$$

The proof is of a purely analytic character. To begin with we introduce Euler's *gamma function* $\Gamma(\tau) := \int_0^\infty t^{\tau-1} e^{-t} dt$.

Exercise. Prove that the function $\Gamma(\tau)$ is defined for all $\tau \geq 0$ and satisfies $\Gamma(\tau + 1) = \tau!$ for each $\tau \in \mathbb{N}$.

(Note that $\Gamma(\tau + 1) = \int_0^\infty t^\tau e^{-t} dt = -t^\tau e^{-t}|_0^\infty + \tau \int_0^\infty t^{\tau-1} e^{-t} dt = \tau \Gamma(\tau)$.)

Lemma 3. *We have*

$$\Gamma(s + 1) = s^{s+1} e^{-s} \int_0^\infty e^{-sf(x)} dx, \text{ where } f(x) = x - \ln x - 1.$$

This follows by making the substitution $t = sx$ in the integral defining $\Gamma(s + 1)$, yielding

$$\Gamma(s + 1) = \int_0^\infty t^s e^{-t} dt = \int_0^\infty s^{s+1} x^s e^{-sx} dx$$

$$= s^{s+1} e^{-s} \int_0^\infty e^{-s(x - \ln x - 1)} dx. \quad \square$$

In view of this lemma, in order to establish Stirling's formula it suffices to show that the integral $I(s) := \int_0^\infty e^{-sf(x)} dx$ is asymptotic to $\sqrt{2\pi/s}$: $I(s) \sim \sqrt{2\pi/s}$, that is, $I(s)\sqrt{s/2\pi} \to 1$ as $s \to \infty$. Of course, the general character of any function asymptotic to $I(s)$ must depend ultimately on the behavior of the function f (whose graph is shown in the the diagram). It turns out that for each s the value of the integral $I(s)$ is already very close to the integral taken over an ε-neighborhood of 1, where the function f is approximately quadratic.

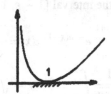

Lemma 4. *One has $\int_0^\infty e^{-x^2} dx = \frac{\sqrt{\pi}}{2}$.*

This can be seen as follows:

$$\left(\int_{\mathbb{R}} e^{-x^2} dx \right)^2 = \int_{\mathbb{R}^2} e^{-x^2 - y^2} dx dy = \int_0^{2\pi} \int_0^\infty r e^{-r^2} dr d\theta$$

$$= \pi \int_0^\infty 2r e^{-r^2} dr = \pi \int_0^\infty e^{-u} du = \pi. \quad \square$$

Corollary. *More generally, $\int_0^\infty e^{-ax^2} dx = \frac{1}{2}\sqrt{\frac{\pi}{a}}$.*

Using this we can prove the following final lemma, from which the theorem is immediate.

Lemma 5. *Let f be any continuous function on $(0, \infty)$ with the following properties:*

(1) *f is decreasing on $(0, 1)$, increasing on $(1, \infty)$, and $f(1) = 0$;*
(2) *there exist numbers $b > 0$ and $c > 1$ such that $f(x) \geq bx$ for all $x \geq c$;*
(3) *there exist positive numbers a, δ, and M such that $f(x) = a(x - 1)^2 + (x - 1)^3 \psi(x)$, where $|\psi(x)| \leq M$ for all $x \in (1 - \delta, 1 + \delta)$.*

Then $I(s) = \int_0^\infty e^{-sf(x)}dx = \sqrt{\pi/(sa)}(1 + o(1))$ as $s \to \infty$.

Exercise. Prove that the function $f(x) = x - \ln x - 1$ has properties (1)–(3).

Here is the proof of Lemma 5. We may clearly assume that δ is sufficiently small that $f(x) \geq a(x - 1)^2/2$ for all $x \in (1 - \delta, 1 + \delta)$, and that s is large enough for $\varepsilon := 1/\sqrt[3]{s} < \delta$ to hold. We break up our integral into a sum of integrals over the subintervals $[0, 1 - \varepsilon]$, $[1 - \varepsilon, 1 + \varepsilon]$, $[1 + \varepsilon, c]$, and $[c, \infty)$. Since $f(x) \geq f(1 - \varepsilon)$ for all $x \in (0, 1 - \varepsilon]$, we have $-sf(x) \leq -sf(1 - \varepsilon) \leq -s\varepsilon^2 a/2 = -a\sqrt[3]{s}/2$, whence

$$\int_0^{1-\varepsilon} e^{-sf(x)}dx \leq \int_0^{1-\varepsilon} e^{-sf(1-\varepsilon)}dx \leq \int_0^{1-\varepsilon} e^{-a\sqrt[3]{s}/2}dx = (1 - \varepsilon)e^{-a\sqrt[3]{s}/2}.$$

Hence the integral over the first subinterval tends to zero exponentially as $s \to \infty$. A similar upper estimate of the integral over the subinterval $[1 + \varepsilon, c]$ shows that that integral likewise tends to zero exponentially as $s \to \infty$. For the fourth integral we have

$$\int_c^\infty e^{-sf(x)}dx \leq \int_c^\infty e^{-sbx}dx = \frac{1}{sb}e^{-sbc},$$

so that this integral also tends to zero exponentially as $s \to \infty$.

Finally, for the integral over the interval $[1 - \varepsilon, 1 + \varepsilon]$ we have

$$\int_{1-\varepsilon}^{1+\varepsilon} e^{-sf(x)}dx \sim \int_{1-\varepsilon}^{1+\varepsilon} e^{-sa(x-1)^2}dx = \frac{1}{\sqrt{sa}} \int_{-\sqrt{sa\varepsilon}}^{\sqrt{sa\varepsilon}} e^{-y^2}dy$$

$$\sim \frac{1}{\sqrt{sa}} \int_{-\infty}^\infty e^{-y^2}dy = \sqrt{\frac{\pi}{sa}}. \quad \square$$

(We leave it to the reader to fill in the details of the above argument.)

Elementary combinatorial problems—such as those in Section 1 of the present chapter—can be used with success in teaching mathematics to students of the middle (i.e., 7th – 9th) grades, since in the course of solving such problems they learn to carry out (and write out) elementary logical arguments, while remaining within the framework of a single formal system. (A vast collection of similar problems may be found in the book [36].) These problems also have the advantage that while on the one hand they can be variously formulated, on the other hand they represent concrete instances of a rather small collection of underlying models. From the teacher's viewpoint they have the not insignificant further advantage that in most cases a student's obtaining the correct answer by accident is as unlikely as his being able to deduce a valid solution from the answer given in advance.

In Section 2 the author provides an approach to introducing the binomial coefficients to students, based on the derivation of the standard recurrence relation between them, shown as arising in the solution of three distinct problems. At the conceptual level, an important role is played by Lemma 1 and its consequences, where the properties of a collection of numerical quantities are investigated using a function somehow associated with the collection. This method is further developed in Section 3 in deriving the formula for the nth Fibonacci number (Theorem 2). We emphasize once again (the teacher *must* appreciate this!) that in order to justify the formal calculation carried out in establishing that formula, one needs to construct a mathematical system appropriate as a context for that calculation; here the most suitable such system is the ring of formal power series. We conclude Section 3 by indicating a different method for obtaining the formula for the Fibonacci numbers, related to difference equations, in turn related to certain linear differential equations.

CHAPTER 3

Geometric Transformations

3.1 Translations, rotations, and other symmetries, in the context of problem-solving

We begin with the following four problems.

Problem 1. Given a straight line ℓ and points A and B off it (in the same plane), find a point (or points) C on ℓ such that the sum of the distances from C to A and B is least.

Problem 2. Given an angle and a point P within it, show how to construct a line segment with midpoint P and endpoints on the arms of the angle.

Problem 3. Two villages are situated on opposite sides of a canal with parallel banks. Where should a bridge be built over the canal so that the path from one village to the other is shortest possible?

Problem 4. Let A, B, C be points of a straight line (with B between A and C), and D and E the vertices of equilateral triangles with bases AB and BC respectively, lying on the same side of the given straight line. Show that the point B together with the midpoints of the segments AE and CD form the vertices of another equilateral triangle.

These problems have elegant solutions involving the ideas of axial and central symmetry, translation, and rotation.

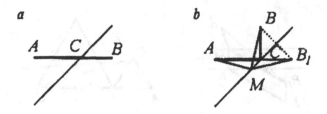

In solving Problem 1, there are two cases to consider. Firstly, if the two points are on opposite sides of the line, then the solution is obvious: the desired point is the point of intersection of the given line with the line segment AB (see Figure a). Suppose, on the other hand, that the points A and B lie on the same side of the line ℓ. In this case denote by B_1 the image point of B under reflection in the line. Then if M is any point on the line (as in Figure b), we have $AM + MB = AM + MB_1$, whence it follows that this sum is least precisely if $M = C$.

Here is one way of solving Problem 2. Let O denote the apex of the given angle and C the point such that M is the midpoint of the segment OC. Through C draw lines parallel to the arms of the given angle, meeting those arms in points A and B, say (see Figure a below). Since $OACB$ is a parallelogram and M is the midpoint of one of its diagonals, M must also be the midpoint of the other diagonal AB.

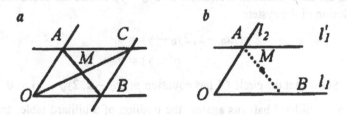

Here is another solution. Denote by ℓ_1, ℓ_2 the arms of the given angle and by ℓ_1' the line symmetrical to ℓ_1 with respect to the point M, i.e., the image of ℓ_1 under reflection in the point M. Let A be the point of intersection of ℓ_2 and ℓ_1' (see Figure b). By construction of ℓ_1', this point is the image of some point B, say, on ℓ_1, under reflection in M. Thus M is the midpoint of the segment AB.

(To thoroughly appreciate the difference between these two solutions, consider Problem 5 below.)

A solution of Problem 3 is depicted below in the figure on the left. The point B_1 is obtained by translating B in a direction perpendicular to the banks of the canal, through a distance equal to the canal's width. Hence the figure BB_1MK is a parallelogram, so that the broken lines $BKMA$ and BB_1MA are equal in length. The bridge should therefore be constructed at the point C.

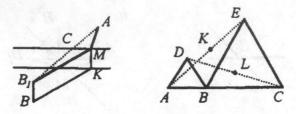

Finally, to Problem 4: Under the counterclockwise rotation of the plane through 60° about the point B, the point C goes to the point E and D to A (refer to the diagram). Hence the segment CD goes to the segment EA, and therefore its midpoint L is sent into the midpoint K of EA. We thus see that the triangle BKL is isosceles with vertical angle 60°, and therefore in fact equilateral.

Here are a few more problems with solutions exploiting the same ideas.

Problem 5. Given two circles and a point (in the plane), construct a line segment having the given point as midpoint and its ends lying on the circles.

The following solution is analogous to the second solution of Problem 2. Denote the given circles by C_1, C_2 and by C_1' the image (also a circle) of the first under reflection in the given point. Then a point of the intersection $C_1' \cap C_2$ will be an endpoint of the desired segment.

Here is an analytic solution: Let $p(x, y) = 0$, $q(x, y) = 0$ be equations of the given circles relative to a Cartesian system of axes, and suppose that $P(x_0, y_0)$ is the given point. If $B(x, y)$ is the endpoint of the desired segment on the second circle, then the other endpoint is $A(2x_0 - x, 2y_0 - y)$. Hence the coordinates x, y form a solution of the system

$$p(2x_0 - x, 2y_0 - y) = 0,$$
$$q(x, y) = 0.$$

Exercise. Show that the circle C_1' has equation $p(2x_0 - x, 2y_0 - y) = 0$.

Problem 6. A billiard ball sits against the cushion of a billiard table. In which direction should the ball be cued so that after bouncing off the three other sides of the table, it returns to its point of departure?

Problem 7. A point P lies on the arc AB of the circle circumscribing an equilateral triangle ABC. Prove that $PC = PA + PB$.

Problem 8. The length of the line segment joining the midpoints of a pair of opposite sides of a quadrilateral is equal to half the sum of the lengths of the other two sides. Prove that the quadrilateral must be a trapezoid.

The last problem has an elegant solution using vectors. The next two problems may be solved using either vector algebra or geometric transformations.

Problem 9. Consider two squares situated in the same plane. Join any corner of one square to any corner of the other by means of a line segment, and then, proceeding in the same direction around both squares, the next corner to the next, and so on. Prove that the midpoints of the four line segments so constructed are also the vertices of a square.

Problem 10. Let $ABCD$ and $DEFG$ be two squares in the same plane, with the labelings of vertices proceeding in the same direction, i.e., both clockwise or both counterclockwise. Prove that the (extended) median through D of the triangle DAG coincides with an altitude of the triangle DCE.

The book [38] contains many elementary geometrical problems with solutions involving motions of the plane.

3.2 Problems involving composition of transformations

Problem 11. Prove that a (nonempty) bounded subset of the plane cannot have more than one center of symmetry.

We use *reductio ad absurdum*. Suppose that A and B are two points such that the corresponding central reflections H_A and H_B of the plane send the given subset to itself. Consider the composite mapping $F := H_B \circ H_A \circ H_B \circ \cdots \circ H_B \circ H_A$ obtained by composing these central symmetries $2n$ times. By assumption, the map F sends the given set to itself. However, the composite map $H_B \circ H_A$ is the translation of the plane through the vector $2\overrightarrow{AB}$ (see the diagram), so that F will be the translation of the plane determined by the vector $2n\overrightarrow{AB}$. Hence for n sufficiently large, the map F will clearly not send the given bounded subset to itself.

A very similar argument shows that a bounded nonempty subset of the plane cannot have two or more distinct parallel axes of symmetry.

Exercise. Prove that the composite of two axial symmetries with intersecting axes of symmetry is a rotation about the point of intersection of these axes through an angle equal to twice the angle between them.

Problem 12. Prove that a closed, convex subset of the plane having two axes of symmetry with angle of intersection incommensurable with π (i.e., not a rational multiple of π), is either a disk or the whole plane.

By the preceding exercise the composite of the symmetries in question is a rotation through an angle incommensurable with π about the point P of intersection of the two axes of symmetry. It follows that the given subset is invariant under rotations Φ_α about that point through angles α forming an everywhere dense subset of the unit circle (see Chapter 7). Since the given subset is assumed to be closed, it must therefore be mapped to itself by every rotation about P. The desired conclusion now follows from the convexity of that subset.

Although the above arguments could not be called complicated, and do not use any so-called "higher mathematics," they are nevertheless of a different sort from those that in high school are usually termed elementary. Here, then, is a problem that is completely elementary in that more usual sense.

Problem 13. Show how to construct a polygon having prescribed points $P_1, P_2, \ldots, P_{2n+1}$ as the midpoints of its sides.

It is not difficult to see that the composite $F := H_{2n+1} \circ H_{2n} \circ \cdots \circ H_1$ of the respective central symmetries with these points as centers is itself a central symmetry about some vertex of the desired polygon (why?). We can find that vertex as follows: taking any point M and its image $M' = F(M)$ under the transformation F, the midpoint of the line segment MM' will be the point we seek. The remaining vertices can then be located without difficulty.

In what follows we shall use the notation Id for the identity map; H_A for the reflection of the plane in the point A (i.e., the central symmetry with center A); R_ℓ for the reflection (axial symmetry) of the plane in the line ℓ; Π_v for the translation determined by the vector v; and $\Phi_\alpha(A)$ for the rotation through the angle α about the point A.

The notation for composition of transformations (or functions generally) is reminiscent of multiplication of numbers. However, composition differs from such multiplication in, for instance, not always being commutative. For example, $H_A \circ H_B = \Pi_{\overrightarrow{2AB}} \neq \Pi_{\overrightarrow{2BA}} = H_B \circ H_A$. It follows that $(H_A \circ H_B) \circ (H_B \circ H_A)$ is the identity transformation. However this is obvious since composition *does* satisfy the associative law, so that

$$(H_A \circ H_B) \circ (H_B \circ H_A) = (H_A \circ (H_B \circ H_B)) \circ H_A = H_A \circ H_A = \text{Id}.$$

We saw above (in Problem 12) that disks are characterized in part by being invariant with respect to an appropriate pair of axial symmetries. In the next two problems we see how, analogously, the relative positions of two geometrical objects may be characterized by some relation between transformations, such as, for instance, commutativity [3]. (For the sake of brevity we shall henceforth omit the circle in composites of transformations.)

Problem 14. Prove the following: 1) $R_{\ell_1} R_{\ell_2} = R_{\ell_2} R_{\ell_1} \iff \ell_1 \perp \ell_2$; 2) $R_\ell H_A = H_A R_\ell \iff A \in \ell$.

Problem 15. Prove the following:

(1) If ℓ_1, ℓ_2, ℓ_3 are distinct straight lines, then $R_{\ell_1} R_{\ell_2} R_{\ell_3} = R_{\ell_3} R_{\ell_2} R_{\ell_1}$ if and only if the three lines are either parallel or concurrent.

(2) $H_A H_B = H_B H_C$ if and only if the point B is the midpoint of the segment AC.

Exercise. What does the relation $H_A H_B H_C H_D = \text{Id}$ imply about the relative positions of the points A, B, C, D?

It is useful in studying motions (geometric transformations) to work out their representations in terms of coordinates with respect to an appropriate Cartesian system. This opens up the possibility of replacing the geometric (i.e., coordinate-free) arguments by computational algebraic ones, although the former are very often both shorter and conducive to a more thorough understanding.

Theorem 1. *In terms of a (rectangular) Cartesian system of coordinates for the plane, the three basic types of Euclidean motions have the following forms:*

(1) *The translation determined by the vector* $v(a, b)$ *has the form*

$$P(x, y) \longmapsto P'(x + a, y + b).$$

(2) *Reflection in the point* (x_0, y_0) *has the form*

$$P(x, y) \longmapsto P'(2x_0 - x, 2y_0 - y).$$

(3) *Rotation about the origin through the angle* α *has the form*

$$P(x, y) \longmapsto P'(x \cos \alpha - y \sin \alpha, \ x \sin \alpha + y \cos \alpha).$$

We shall derive only the third of these formulae. Let r denote the length of OP and φ the angle between the x-axis and the vector $\overrightarrow{OP}$. Then $OP' = OP = r$, and the angle between the x-axis and the vector $\overrightarrow{OP'}$ is $\varphi + \alpha$ (see the diagram). Hence the x-coordinate of the point P' is $r \cos(\varphi + \alpha) = r \cos \varphi \cos \alpha - r \sin \varphi \sin \alpha = x \cos \alpha - y \sin \alpha$. The y-coordinate of P' can be found similarly. $\square$

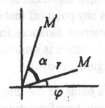

Exercise. Rederive the formula of part (3) of the theorem without using the sum rule for the cosine (or sine), and then deduce the latter from that formula.

Exercise. Prove that the composite of two rotations is either again a rotation or else a translation.

3.3 The group of Euclidean motions of the plane

So far, the only geometric transformations we have used are reflections in a point or line, translations, and rotations. We remind the reader that a transformation F of the plane (a map of $\mathbb{R}^2$ into itself) is called a *(Euclidean) motion* if it preserves

(Euclidean) length between points, i.e., if for all points A, B, we have $AB = A'B'$ where $A' := F(A)$, $B' := F(B)$. Clearly, all of the transformations considered so far in this chapter are motions.

The following lemma can be verified without difficulty.

Lemma 1. *(a) Every motion of the plane is bijective, and therefore has an inverse transformation. (b) The inverse of any motion and the composite of any two motions are again motions.* □

Thus the set of all motions of the plane furnished with the operation of composition, is a (nonabelian) group (the "plane Euclidean group").

Lemma 2. *If a motion of the plane fixes each of two distinct points, then it is either the identity transformation or the reflection in the straight line through those two points.*

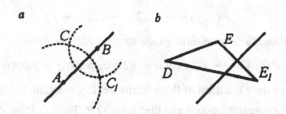

Denote by A and B the two fixed points. Let C be any point on the line segment AB and denote by C_1 the image of C under the motion in question. Since $AC_1 + C_1B = AC + CB = AB$, the point C_1 must also lie on the segment AB, whence it follows that in fact $C_1 = C$. A similar argument shows that every point of the line through A and B is fixed. If C is any point off that line, then it is easy to deduce from the fact that the motion preserves distances from the fixed points A and B, that either C is also fixed or that its image is symmetrically placed with respect to the line through A and B (see Figure a above).

Exercise. Prove that if there is a point off the line through A and B that is also fixed, then the motion is the identity transformation.

(Consult Figure b above: $DE_1 \geq DE$.) □

Corollary. *The only motion of the plane having three noncollinear fixed points is the identity transformation.*

Theorem 2. *Every motion of the plane can be expressed as a composite of three or fewer axial symmetries (reflections in lines).*

Let F be any motion and A, B, C any three noncollinear points of the plane. Denote the images of A, B, C under F by A_1, B_1, C_1 respectively. If $A \neq A_1$, write R_1 for the reflection in the perpendicular bisector of the segment AA_1; the motion R_1 then sends A_1 back to A. Write B_2, C_2 respectively for the images of B, C under R_1. (If $A_1 = A$, we take the identity transformation in place of

R_1, in which case $B_2 = B_1, C_2 = C_1$; i.e., we effectively omit this first step.)
Proceeding to the next step, if $B_2 \neq B$, then the reflection R_2; say, in the bisector
of the angle $\angle BAB_2$, sends B_2 to B. Denote by C_3 the image of the point C_2 under
this reflection. (If $B_2 = B$, take the identity transformation in place of R_2.) If
$C_3 = C$, then the composite motion $R_2 R_1 F$ has three noncollinear fixed points,
whence by the above corollary $R_2 R_1 F = \text{Id}$, giving $F = R_1 R_2$. If $C_3 \neq C$, then
$R_2 R_1 F = R_3$, the reflection in the line through A and B, whence $F = R_1 R_2 R_3$.
□

Corollary. *The axial symmetries form a set of generators for the group of all motions of the plane. Hence the plane Euclidean group has a generating set consisting
of elements of order 2.*

For any subset of the plane we define the *group of symmetries* of that subset to
be the set of all motions sending that subset onto itself.

Exercise. Prove that such a set of motions is indeed always a group (a subgroup
of the Euclidean group).

For example, the symmetry group of an equilateral triangle consists of 6 motions;
it is isomorphic to the group S_3 of all permutations of three symbols. This follows
most easily from the fact that any plane motion is determined by its effect on any
three noncollinear points (see the last corollary but one). In view of this fact the
symmetries of the triangle are determined by how they map its vertices. Since the
images of the vertices must again be vertices, and since every permutation of the
vertices can be realized by some symmetry, the claim is established. (As generators
of this group we may take the reflections in any two medians.)

Exercise. Prove that the symmetry group of an equilateral triangle is isomorphic
to the abstract group with two generators a, b of order two satisfying the additional
relation $aba = bab$.

Exercise. Find a subset of the plane whose symmetry group is isomorphic to the
group of integers $(\mathbb{Z}, +)$.

The next problem asks for a classification of plane motions in terms of
coordinates referred to some rectangular Cartesian system of axes.

Problem 16. Find necessary and sufficient conditions on the numbers a, b, c, d
for the map of the plane given by $(x, y) \longmapsto (ax + by, cx + dy)$ to be a motion.

3.4 Ornaments

In this section we apply the algebraic approach via symmetry groups to the problem
of classifying "frieze patterns," or "ornaments," i.e., patterns along a band. The two
ornaments shown in the diagram are, as the man in the street might put it, symmetric
in distinct ways; a mathematician would express the idea more precisely by saying
simply that they have different symmetry groups.

We define an *ornament* to be a closed subset of the plane lying in the strip between two parallel lines with the property that its symmetry group G contains some nontrivial translation, but not all translations, parallel to the strip. We shall assume the strip has least width subject to containing the ornament.

Lemma 3. *For any ornament $\mathcal{M}$, the set of all translations contained in the symmetry group of $\mathcal{M}$ is a subgroup isomorphic to $\mathbb{Z}$; i.e., there is a translation (which will be denoted by Π in the sequel) with the property that every other translation sending $\mathcal{M}$ to itself is some power Π^k, $k \in \mathbb{Z}$.*

Let Π_v be any nontrivial translation in the symmetry group G of the ornament. Consider the subset M of the real line defined by $M := \{\lambda \in \mathbb{R} \mid \Pi_{\lambda v} \in G\}$. Clearly, M is a nontrivial, proper, closed subgroup of $(\mathbb{R}, +)$. Hence (see Chapter 7) M has the form $\{k\alpha \mid k \in \mathbb{Z}\}$ for some fixed number α, and we may take $\Pi = \Pi_{\alpha v}$. □

We shall call the translation Π the *period* of the given ornament, and the infinite cyclic subgroup C_∞ it generates the *subgroup of periods*.

Theorem 3. *As above, let G and C_∞ denote respectively the symmetry group and subgroup of periods of any ornament. There are precisely six distinct (i.e., nonisomorphic) pairs (G, C_∞), and seven distinct geometrical realizations of these group–subgroup pairs.*

We shall establish this theorem by carefully sorting through all possible variants. For this purpose the following auxiliary result will be useful. (We leave its proof to the reader as an exercise.) The straight line along the middle of the strip containing an ornament will be called the *axis* of the ornament.

Lemma 4. *Apart from translations, the symmetry group of an ornament can contain only symmetries of the following types:*

(1) *At most one axial symmetry (reflection in a line) about the axis of the ornament (denoted by R').*
(2) *Axial symmetries about lines perpendicular to the strip containing the ornament.*
(3) *Central symmetries about points on the axis of the ornament.*
(4) *At most one glide reflection (i.e., the composite of a reflection in a line and a translation in the direction of that line) about the axis of the ornament (denoted by U).* □

To prove Theorem 3, we first introduce a convenient coordinate system: as x-axis we take the axis of the ornament, and arrange the scale so that the translation Π moves the plane through the vector $(1, 0)$. Denote by R_a and H_a, respectively, the reflection in the line $x = a$ and the central symmetry about the point $(a, 0)$.

Case 1. If the symmetry group contains only translations, then $G = C_\infty = \{\Pi^k | k \in \mathbb{Z}\}$.

Case 2. Suppose that G contains the reflection R' in the axis of the ornament, and is generated by this element together with C_∞. Since R' and Π commute, the group G is isomorphic to an abelian group with two generators, one of order 2 and the other of infinite order. It follows that $G \cong \{a, b | b^2 = 1\} \cong \mathbb{Z} \times \mathbb{Z}_2$.

Case 3. Suppose that G is generated by Π (inevitably!) together with an axial symmetry R_0 about a line perpendicular to the axis of the ornament. Choose the origin of coordinates so that the latter line is the y-axis. The composite symmetry $R_0 \Pi$ then sends an arbitary point (x, y) to $(-x - 1, y)$, as does $\Pi^{-1} R_0$; hence $R_0 \Pi = \Pi^{-1} R_0$. Thus G has generators a and c satisfying the relations $a^2 = 1$ and $ac = c^{-1}a$, or equivalently, $a = cac$ or (since $a^2 = 1$) $acac = 1$. Writing $b := ac$, it follows that $G = \{a, b | a^2 = b^2 = 1\}$. (Show that there are no relations that are not consequences of the two exhibited!)

This group, the "infinite dihedral group," is usually denoted by the symbol D_∞. Observe that the groups $\mathbb{Z} \times \mathbb{Z}_2$ and D_∞ both contain the subgroup corresponding to C_∞ as a subgroup of index 2; however, since the second of these is nonabelian, we certainly have $D_\infty \not\cong \mathbb{Z} \times \mathbb{Z}_2$.

Exercise. For each real number α, denote by R_α the reflection in the line $x = \alpha$. Show that in the present case (Case 3) $R_{k/2} \in G$ for every integer k.

(Note that if $R_\alpha \in G$, then since $R_\alpha R_0$ is the translation through the distance 2α, we must have $2\alpha \in \mathbb{Z}$.)

Case 4. Suppose G is generated by H_0 and Π. The group of the ornament is in this case again isomorphic to D_∞.

Case 5. Suppose G is generated by Π together with a glide reflection U' : $(x, y) \longmapsto (x + u, -y)$. Since $(U')^2 = \Pi_{2u}$, we have $2u \in \mathbb{Z}$; write $k = 2u$. If $u \in \mathbb{Z}$, then $\Pi^{-u} U' = R'$, so that $R' \in G$, and it follows that in fact G is generated by Π and R', which is Case 2 above. On the other hand if $2u = 2l + 1$, i.e., if u is not an integer, then the composite $\Pi^{-l} U'$ is the glide reflection $U_{1/2}$, which we shall denote by U. Then since $\Pi = U^2$, we have $G = \{U^k | k \in \mathbb{Z}\} \cong \mathbb{Z}$, and here G contains C_∞ as the subgroup consisting of the even powers of the generator U.

Case 6. Next suppose G is generated by R', R_0, and Π. Essentially since the symmetry R_0 commutes with R' (their respective axes of symmetry being perpendicular) and also with Π, it follows that $G \cong D_\infty \times \mathbb{Z}_2$. Note that in this case since $H_0 = R' R_0$, if G contains any two of the symmetries H_0, R', R_0, then it will also contain the third.

The situation where $R', U \in G$ is not possible, since $R'U = \Pi_{1/2}$, and this translation is not a symmetry of the ornament.

Case 7. Finally, suppose G is generated by H_0 and U. Then

$$(x, y) \xmapsto{U} \left(x + \frac{1}{2}, -y\right) \xmapsto{H_0} \left(-x - \frac{1}{2}, y\right) \xmapsto{U} (-x, -y),$$

i.e., $U H_0 U = H_0$. It follows that G is isomorphic to the abstract group with two generators a, b and defining relations $a^2 = 1$, $bab = a$ (or, equivalently, $(ab)^2 = 1$), whence $G \cong D_\infty$.

Exercise. Prove that the pair (G, C_∞) of this case is not isomorphic to either of the corresponding pairs of cases 3 and 4.

Exercise. Show that if $H_0, U \in G$, then $R_{\frac{1}{4}} \in G$, and if $R_0, U \in G$, then $H_{\frac{1}{4}} \in G$.

In effect the only case not covered above is that where G is generated by H_0 and R_α for some real number α. Here since $R_\alpha H_0 : (x, y) \longmapsto (2\alpha + x, y)$, we have $R_\alpha H_0 = U_{2\alpha}$, whence $4\alpha \in \mathbb{Z}$. It is now not difficult to see that this case reduces to the preceding one.

Exercise. Draw seven ornaments representing the seven cases analyzed above. To which of these cases belong the two ornaments depicted in the diagram at the beginning of this section?

3.5   Mosaics and discrete groups of motions

Problem 17. Show that the plane can be tiled (covered without overlaps or gaps) with tiles all congruent to an arbitrary given plane quadrilateral (not necessarily convex).

First consider the parallelogram with vertices the midpoints of the sides of the given quadrilateral; denote these vertices by E_{00}, E_{01}, E_{11}, and E_{10} (see the diagram below). Coordinatize the plane using the coordinate system with origin E_{00} and basis vectors $\mathbf{i} := \overrightarrow{E_{00}E_{10}}$ and $\mathbf{j} := \overrightarrow{E_{00}E_{01}}$. Write H_{kl} for the reflection of the plane in the point with coordinates (k, l) with respect to this system. We claim that the plane is tiled by the images of the given quadrilateral under all motions H_{kl}. If two distinct such images had an interior point in common, then the composite $H_{mn}H_{kl}$ of the corresponding reflections would send some interior point of the original quadrilateral to another such point, and this is not possible, since this composite motion is just the translation of the plane through a vector of the form $2(a\mathbf{i} + b\mathbf{j})$, $a, b \in \mathbb{Z}$. Finally, since by means of a translation of this form every point of the plane can be mapped to a point either of the original quadrilateral or of one or the other of two adjacent images (see the diagram), it follows that the image quadrilaterals do indeed cover the plane.

It is not difficult to see that provided the given quadrilateral is sufficiently irregular, the set of all motions of the plane sending the corresponding mosaic (or tesselation) to itself is precisely the group consisting of all composites of arbitrarily many of the reflections in the points of the lattice of parallelograms determined by the vectors **i**, **j**, i.e., the group generated by these reflections. This group, denoted by **p2** in the standard notation, is one of the 17 possible discrete groups of isometries of the Euclidean plane containing two linearly independent translations. Here we shall only touch on this classification. (It is interesting that every one of these groups is, in effect, represented by the mosaics with which in the thirteenth century the moors decorated the walls of the Alhambra, the celebrated palace of Moorish sultans in Granada.)

By a *regular system of points* of the plane [15] we shall mean a subset of points with the following three properties: (1) every bounded region of the plane should contain only finitely many points of the subset; (2) the number of points of the subset in any disk should be proportional to the square of the radius of the disk; (3) for each pair of points of the subset there should exist a symmetry of the subset sending one of the two points to the other.

We define a "general lattice" in the plane as follows: Let **a** and **b** be two linearly independent vectors in the plane, of different lengths and with the angle between them not equal to $\pi/3$, $\pi/2$, or $2\pi/3$. Choose any point E of the plane. The *general lattice* determined by **a** and **b** (and E) is then defined to be the (vector) lattice consisting of all points E_{kl} such that $\overrightarrow{E E_{kl}} = k\mathbf{a} + l\mathbf{b}$ for some $k, l \in \mathbb{Z}$.

Exercise. Prove that such a lattice is symmetric about the midpoint of the line segment joining any two of its points.

(The totality of such midpoints also forms a vector lattice, determined by the vectors **a**/2 and **b**/2.)

Lemma 5. *The symmetry group of a general lattice consists of the translations through all vectors of the form $k\mathbf{a} + l\mathbf{b}$, $k, l \in \mathbb{Z}$, together with the reflections in the points of the double refinement of the lattice.*

That the symmetry group contains these translations is obvious. That it contains these reflections is given to us by the preceding exercise. That it contains no other motions follows from the condition on the angle between **a** and **b**. $\square$

Lemma 6 ([3]). *The only possible finite orders of elements of the symmetry group of a regular system of points in the plane, are $2, 3, 4$ and 6.*

Exercise. Prove that a plane motion of finite order $n > 2$ must be a rotation about some point through an angle $2\pi k/n$ where k and n are relatively prime.

Hence a symmetry S of finite order $n > 2$ of the given regular point–system is a rotation through $2\pi k/n$ where k and n are relatively prime. Therefore, there exist integers a, b satisfying $ak + bn = 1$, whence $2\pi ak/n + 2\pi b = 2\pi/n$, so that $S' := S^a$ is a rotation through the angle $2\pi/n$. Let A denote the center of rotation of S'. It follows from condition (3) of the definition of a regular point–system that every point of the system is the center of a rotation of order n which is a symmetry of the system. Let B be a closest point of the system to A. (That there *is* a closest point is guaranteed by condition (1).) Let A' denote the image of A under any symmetry rotating the plane through $2\pi/n$ about B, and B' the image of B under a symmetry rotating the plane through the same angle about A' (see the diagram). If $A = B'$, then $n = 6$. Otherwise, since B is a closest point of the system to A, we must have $AB' \geq AB$, whence $n \leq 4$. $\square$

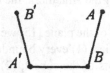

Exercise. Prove that if a rotation of the plane is a symmetry of a regular system of points, then it must be a rotation through an angle of the form $2\pi k/n$, i.e., $S^n = \text{Id}$.

The material of this chapter naturally falls into two (unequal) parts. The first of these consists of the elementary problems of plane Euclidean geometry occupying the first section, interesting in their own right. It is instructive to compare the two methods of solution of these problems, the one using basic transformations of the plane and the other analytic geometry (involving rectangular Cartesian coordinate systems and vector algebra). For instance, the basic idea underlying the geometric solutions of Problems 2 and 5 may be inferred from the given analytic solution of the second of these problems.

Of course, the most elegant solutions of the problems of Section 3.1, are those obtained using the method to which that section is devoted. However, it seems to be impossible to teach high-school students to solve geometrical problems using this approach; what is important here is the inculcation of taste for the sort of argumentation involved, and, what is absolutely crucial, accustoming the students to thinking of mappings (motions, functions) as mathematical objects on the same footing as, for instance, the real numbers.

The whole of the rest of this chapter is in one way or another concerned with composition of mappings. This concept makes its appearance in the traditional high-school mathematics program in the guise of functions of some degree of complexity. However, in that context the emphasis is mainly on the concrete computation of the values of such functions, and the idea of regarding the function (given via one or another complicated formula) "as a whole," so to speak, is left out of account. Section 2 begins with a simple yet engaging problem for whose solution just this "holistic" point of view turns out to be expedient; Problems 14 and 15 are even more typical in this respect. From the point of view of methodology, this section provides a bridge from the elementary material to that of Sections 4 and 5, where symmetry groups are considered. Note that computation of composites of transformations may seem less forbidding to the students if the transformations are first given coordinate form (as in Theorem 1), and then the composite calculated in terms of those coordinates (as in the proof of Theorem 3).

CHAPTER 4

Inequalities

4.1 The means of a pair of numbers

The theme of this chapter is vast, and the author has here intentionally limited himself (and the reader) to the classical inequalities [12].

We begin with problems using one of the best-known of inequalities, namely that between the arithmetic and geometric means of two nonnegative real numbers (the a.m.-g.m. inequality):

$$\frac{a+b}{2} \geq \sqrt{ab}.$$

Problem 1. Prove the following inequalities:

(1) $|a/b + b/a| \geq 2$.

(2) $|x + a/x| \geq 2\sqrt{a}$, $a \geq 0$.

(3) $\sqrt{a+b} \leq \sqrt{a} + \sqrt{b} \leq \sqrt{2(a+b)}$, $a, b \geq 0$.

(4) $\log_2 \pi + \log_\pi 2 > 2$.

(5) $a^2 + b^2 + c^2 \geq ab + bc + ca$.

(6) $(1 + a_1)(1 + a_2)\cdots(1 + a_n) \geq 2^n$, if $a_i > 0$ and $a_1 a_2 \cdots a_n = 1$.

(7) $n! \leq ((n+1)/2)^n$.

(8) $n! \leq n^n 2^{(1-n)}$.

(9) $(n!)^2 \geq n^n$.

(10) $(a^2 + b^2)/(a + b) \leq (a^3 + b^3)/(a^2 + b^2)$, $a, b > 0$.

(11) $x^2 + y^2 + z^2 \geq \frac{1}{3}(x + y + z)^2$.

(12) $a/(b + c) + b/(a + c) + c/(a + b) \geq 3/2$, $a, b, c > 0$.

(13) $(a + b)(b + c)(c + a) \geq 8abc$, $a, b, c > 0$.

(14) $abc \geq (a + b - c)(b + c - a)(c + a - b)$, where a, b, c are the lengths of the sides of a triangle.

(15) $(a + b)^4 \leq 8(a^4 + b^4)$.

(16) $a + b + c + d \geq 4\sqrt[4]{abcd}$.

These inequalities are closely interrelated. We shall prove only (12).

Setting $u := b+c$, $v := a+c$, $w := a+b$, we have $a = (v+w-u)/2$, $b = (u+w-v)/2$, and $c = (u+v-w)/2$. Substituting these expressions for a, b, c in (12) yields the equivalent inequality

$$\frac{v+w-u}{2u} + \frac{u+w-v}{2v} + \frac{u+v-w}{2w} \geq \frac{3}{2},$$

which simplifies to the obvious inequality

$$\left(\frac{v}{u} + \frac{u}{v}\right) + \left(\frac{w}{u} + \frac{u}{w}\right) + \left(\frac{v}{w} + \frac{w}{v}\right) \geq 6.$$

Next, three problems at first glance of a geometric character.

Problem 2. A unit square is subdivided into four rectangles by means of two line segments parallel to its sides. Show that the product of the areas of two nonadjacent rectangles is not more than $1/16$.

Problem 3. Let s_1, s_2 denote the areas of the two parts into which a straight line divides a regular hexagon of side 1. Find the largest possible value of the product $s_1 s_2$.

(Since the sum $s = s_1 + s_2$ is equal to the area of the hexagon, it follows that $s_1 s_2 \leq s^2/4 = 27/16$.)

Both of these problems are solved using the inequality $x(a - x) \leq a^2/4$, which of course follows in turn from the fact that the square of a real number is never negative.

Problem 4. Find the image of the plane under the map $f(x, y) := (x + y, xy)$, and explain your answer algebraically.

Problem 5. Find the lengths of the edges of a rectangular parallelepiped (box with rectangular faces) (a) of least surface area among all such parallelepipeds with main diagonal of unit length; (b) having the longest main diagonal among all such parallelepipeds with unit surface area.

(Both of these problems are solved using inequality (5) of Problem 1.)

Problem 6. Find the largest value of the function $f(x, y) := xy$ on the set of points satisfying the inequality $x^2 + xy + y^2 \leq 1$.

We conclude this subcollection with inequalities between various means of two positive numbers, and the geometric interpretations of these means.

Problem 7. Establish the following inequalities for any positive real numbers a, b:

$$\frac{2}{a^{-1} + b^{-1}} \leq \sqrt{ab} \leq \frac{a+b}{2} \leq \sqrt{\frac{a^2 + b^2}{2}} \leq \max\{a, b\}.$$

Problem 8. Prove that in a trapezoid with parallel sides (bases) of lengths a, b:

(1) $\sqrt{\dfrac{a^2 + b^2}{2}}$ is the length of the line segment parallel to the bases, dividing the trapezoid into two subtrapezoids of equal area.

(2) $\dfrac{a+b}{2}$ is the length of the line segment through the middle of the trapezoid, i.e., parallel to the bases and midway between them.

(3) $\dfrac{2ab}{a+b}$ is the length of the line segment parallel to the bases of the trapezoid and passing through the point of intersection of its diagonals.

(4) $\sqrt{ab}$ is the length of the line segment subdividing the trapezoid into two similar ones.

Many problems have solutions using the inequality (due to Cauchy) between the arithmetic and geometric means in the particular case of three (nonnegative) numbers.

Problem 9. (a) Prove that if $a + b + c = 0$, then $a^3 + b^3 + c^3 = 3abc$.

(b) Factor $a^3 + b^3 + c^3 - 3abc$.

(c) Deduce the inequality $\dfrac{x+y+z}{3} \geq \sqrt[3]{xyz}$.

Part (a) of this problem gives a hint for part (b), since it suggests looking for a factorization of the polynomial $a^3+b^3+c^3-3abc$ in the form $(a+b+c)P_2(a, b, c)$, where P_2 is a homogeneous polynomial of degree 2, i.e., $P_2(a, b, c) = a^2 + b^2 + c^2 + k(ab + bc + ca)$. It is not difficult to see that the only possibility is $k = -1$, and then a direct verification shows that indeed $a^3 + b^3 + c^3 - 3abc = (a + b + c)(a^2 + b^2 + c^2 - ab - bc - ca)$.

For part (c), set $x := a^3$, $y := b^3$, $z := c^3$, and use the factorization of part (b) together with inequality (5) of Problem 1.

Problem 10. Prove that:

(a) The least value taken by the function $y = ax^2 + 2b/x$ on the set of all positive x is $3\sqrt[3]{ab^2}$.

(b) The greatest volume of an open box made from a square sheet of cardboard of side a with square pieces cut out of its corners is $2a^3/27$.

(c) Among all rectangular parallelepipeds of unit volume, the cube has least surface area.

Solutions: (a) We have

$$ax^2 + \frac{2b}{x} = ax^2 + \frac{b}{x} + \frac{b}{x} \geq 3\sqrt[3]{ax^2 \frac{b}{x} \frac{b}{x}} = 3\sqrt[3]{ab^2}.$$

(b) Denoting by x the length of a side of any of the square corner pieces cut out of the sheet of cardboard, we have

$$V = x(a - 2x)^2 = \frac{1}{4}(4x)(a - 2x)(a - 2x)$$

$$\leq \frac{1}{4}\left(\frac{4x + a - 2x + a - 2x}{3}\right)^3 = \frac{2a^3}{27}.$$

(c) We have $ab + bc + ca \geq 3\sqrt[3]{a^2b^2c^2} = 3$.

Observe that the above solution of (a) is in fact not quite complete, since, for instance, although $ax^2 + 2b/x = ax^2 + b/(2x) + 3b/(2x) \geq 3\sqrt[3]{3ab^2/4}$, the final expression here is not the least *value* of the given function. Note also that instead of using the a.m.–g.m. inequality of Cauchy, we could of course have used the derivative of the function to find its least value.

The a.m.–g.m. inequality has some rather unexpected applications.

Problem 11. Find all integer solutions of the equation

$$\frac{xy}{z} + \frac{yz}{x} + \frac{zx}{y} = 3.$$

It is also useful for solving certain geometrical problems.

Problem 12. Prove that among all triangles of a fixed perimeter, the equilateral triangle has (a) the circumscribed circle of least radius; (b) the inscribed circle of greatest radius.

For the radius r of the inscribed circle we have, writing A for the area of the triangle,

$$r = \frac{2A}{a + b + c} = \frac{1}{2}\sqrt{\frac{(a + b - c)(b + c - a)(c + a - b)}{a + b + c}}$$

$$\leq \frac{1}{2}\sqrt{\frac{abc}{a + b + c}} \leq \frac{1}{2}\sqrt{\frac{(a + b + c)^2}{27}}.$$

In the following problem, in contrast to the usual sort of problem where some quantity is to be maximized or minimized over a class of geometrical entities, it is not at all simple to guess the answer.

Problem 13. What are the lengths of the edges of a rectangular parallelepiped of unit volume ensuring that the length of the shortest path along its surface from any vertex to the opposite one is least?

Let a, b, c denote the lengths of the edges of the parallelepiped. By developing (i.e., unfolding) the parallelepiped onto a plane, one sees readily that the length d, say, of the shortest path along the surface of the parallelepiped joining a pair of opposite vertices is the least of the quantities $d_1 = \sqrt{a^2 + (b + c)^2}$, $d_2 = \sqrt{b^2 + (c + a)^2}$, $d_3 = \sqrt{c^2 + (b + a)^2}$. Since $abc = 1$, one has

$$a^2 + (b + c)^2 = \frac{1}{b^2c^2} + (b + c)^2 \geq \frac{1}{b^2c^2} + 4bc \geq 3\sqrt[3]{4},$$

with equality precisely when $b = c = 1/\sqrt[6]{2}$, $a = \sqrt[3]{2}$. It then remains only to observe that with these values of a, b, c one has $d_1 = \min\{d_1, d_2, d_3\}$.

We conclude this introductory section with the following simple, although somewhat unusual, problem.

Problem 14. Prove that for positive a, b, c the inequalities
$a(1 - b) > 1/4$, $b(1 - c) > 1/4$, $c(1 - a) > 1/4$ cannot hold simultaneously.

(Deduce from the first that $a - b > 0$, and so on.)

4.2   Cauchy's inequality and the a.m.–g.m. inequality

We begin our exposition of the classical inequalities with Cauchy's (or the Cauchy–Bunyakovskiĭ) inequality.

Theorem 1 (Cauchy's inequality).   *For every n real numbers $a_1, a_2, \ldots, a_n$ the following inequality holds:*

$$\left(\sum_{i=1}^{n} a_i b_i \right)^2 \leq \sum_{i=1}^{n} a_i^2 \sum_{i=1}^{n} b_i^2.$$

Consider the following quadratic function of t: $\Phi(t) := \sum_{i=1}^{n}(a_i - b_i t)^2$. We have $\Phi(t) \geq 0$ for all $t \in \mathbb{R}$, with $\Phi(t) = 0$ for some value of t if and only if the n–tuples $(a_1, \ldots, a_n)$ and $(b_1, \ldots, b_n)$ are proportional. Rewriting $\Phi(t)$ explicitly as a quadratic polynomial in t, we obtain

$$\Phi(t) = t^2 \sum_{1}^{n} b_i^2 - 2t \sum_{1}^{n} a_i b_i + \sum_{1}^{n} a_i^2 \geq 0,$$

whence $(\sum_{1}^{n} a_i b_i)^2 \leq \sum_{1}^{n} a_i^2 \sum_{1}^{n} b_i^2$, with equality if and only if $(a_1, \ldots, a_n)$ and $(b_1, \ldots, b_n)$ are proportional. $\square$

Problem 15. Prove the following inequalities:
  (1) $|a \cos x + b \sin x| \leq \sqrt{a^2 + b^2}$.
  (2) $\sum_{i=1}^{n} a_i \sum_{i=1}^{n} a_i^{-1} \geq n^2$, $a_i > 0$.
  (3) $\sum_{i=1}^{n} a_i^2 \geq 1/n$, if $\sum_{i=1}^{n} a_i = 1$, $a_i > 0$.
  (4) $\sqrt{\sum_{i=1}^{n}(x_i + y_i)^2} \leq \sqrt{\sum_{i=1}^{n} x_i^2} + \sqrt{\sum_{i=1}^{n} y_i^2}$. (When do we have equality here?)

The last inequality in this problem, and Cauchy's inequality itself, have the following well–known geometric interpretations: For any vectors $\mathbf{a}, \mathbf{b} \in$ Euclidean $\mathbb{R}^n$,

$$|\mathbf{a} + \mathbf{b}| \leq |\mathbf{a}| + |\mathbf{b}|, \text{ and } |\mathbf{a} \cdot \mathbf{b}|^2 \leq |\mathbf{a}|^2 |\mathbf{b}|^2.$$

See below for generalizations of these.

Problem 16. (a) Prove that for every natural number n and real number $h > -1$, one has the inequality $(1 + h)^n \geq 1 + nh$ (often attributed to Jacques Bernoulli, but actually known earlier).

(b) Methuselah Smith deposits one dollar in the bank at 1% (compound) interest per annum. Prove that after 1000 years there will be at least $1000 in his account.

(To prove Bernoulli's inequality is a simple exercise in mathematical induction. Part (b) is established using that inequality—but not in one step!)

Here at last is the much–touted a.m.–g.m. inequality (due to Cauchy) in full generality:

Theorem 2. *For any n nonnegative real numbers* $a_1, \ldots, a_n$ *we have*

$$\sqrt[n]{\prod_{i=1}^{n} a_i} \le \frac{1}{n} \sum_{i=1}^{n} a_i, \ a_i \ge 0.$$

In view of the importance of this inequality, we shall give five different proofs of it. We leave the first step of the first proof to the reader.

Exercise. Prove the a.m.–g.m. inequality in the case where n is a power of 2.

Now suppose that n is arbitrary. Let k be a natural number sufficiently large that $n < 2^k =: m$. Set $\xi := \frac{1}{n} \sum_{i=1}^{n} a_i$, and consider the ordered m–tuple of numbers $b_1, b_2, \ldots, b_m$, where $b_i = a_i$ for $i \le n$ and $b_i = \xi$ for $n + 1 \le i \le m$. By the exercise we have $\sqrt[m]{\prod_{l=1}^{m} b_l} \le \frac{1}{m} \sum_{i=1}^{m} b_i$, which is equivalent to the inequality

$$\left(\prod_{i=1}^{n} a_i \cdot \xi^{n-m} \right)^{1/m} \le \frac{1}{m} \left(\sum_{i=1}^{m} a_i + (m - n)\xi \right).$$

Since the right-hand side is just $\frac{1}{m}(n\xi + (m - n)\xi) = \xi$, a little algebraic manipulation yields $\prod_{i=1}^{n} a_i \le \xi^n$. □

Our second proof begins with the following argument. We introduce the notation

$$\Pi(a_1, \ldots, a_n) := \sqrt[n]{\prod_{i=1}^{n} a_i}.$$

Suppose that not all of the numbers $a_1, \ldots, a_n$ are equal; then by re–indexing them we may suppose that $a_1 \ne a_2$. This assumed, we have $a_1 a_2 < (a_1 + a_2)^2/4$, whence

$$\Pi(a_1, \ldots, a_n) < \Pi\left(\frac{a_1 + a_2}{2}, \frac{a_1 + a_2}{2}, a_3, \ldots, a_n \right).$$

Hence for a fixed value ξ of the arithmetic mean $\frac{1}{n} \sum_{i=1}^{n} a_i$ (considered as a function of variables $a_1, \ldots, a_n$), the largest value of the function Π is attained only at inputs $a_1^*, \ldots, a_n^*$ satisfying $a_1^* = \cdots = a_n^*$, at which we have $\Pi(a_1^*, \ldots, a_n^*) = a_1^* = \xi$.

This argument has the pedagogical advantage of having a gap. What it *does* show is that if the function Π *has* a largest value (subject to $\frac{1}{n} \sum_{i=1}^{n} a_i$ remaining fixed at the value ξ), then this is attained when the n inputs are all equal. However it is in general entirely possible for the supremum of all admissible values of a function not actually to be reached, i.e., not to be a value of the function. However, since in the present situation the function Π is continuous and the subset $\{(a_i)_{i=1}^n \mid a_i \geq 0, \sum_1^n a_i \leq n\xi\}$ of $\mathbb{R}^n$ is compact, we are assured by Weierstrass's theorem that it does attain a greatest value on that subset. $\square$

The third proof is at first glance surprising, since it reduces the a.m.–g.m. inequality to an inequality in a single variable. We argue by induction. Jumping to the inductive step, we suppose that

$$\sqrt[n-1]{\prod_{i=1}^{n-1} a_i} \leq \frac{1}{n-1} \sum_{i=1}^{n-1} a_i.$$

This implies that $\sum_1^n a_i \geq (n-1)\sqrt[n-1]{\prod_1^{n-1} a_i} + a_n$, so that it suffices to show that $(n-1)\sqrt[n-1]{\prod_1^{n-1} a_i} + a_n \geq n\sqrt[n]{\prod_1^n a_i}$. If we define θ by $\theta^{n(n-1)} := \prod_1^{n-1}(a_i/a_n)$, then in terms of θ the last inequality becomes $(n-1)\theta^n + 1 \geq n\theta^{n-1}$, or $n\theta^{n-1}(\theta - 1) \geq \theta^n - 1$, and this inequality is easy to establish. $\square$

Note that the last inequality has the form $f(\theta) - f(1) \leq f'(\theta)(\theta - 1)$, where $f(x) := x^n$, whence we see that it follows from the mean–value theorem together with the fact that $f'(\theta)$ is increasing; thus its geometric import is simply that the graph of the function f is concave up.

The fourth proof uses the following lemma.

Lemma 1. *Let* a_i, $q_i > 0$, $i = 1, 2, \ldots, n$, $\sum_{i=1}^n q_i = 1$. *Then*

$$\lim_{x \to 0} \ln \left(\sum_{i=1}^n q_i a_i^x \right)^{1/x} = \sum_{i=1}^n q_i \ln a_i.$$

This is shown as follows:

$$\lim_{x \to 0} \ln \left(\sum_{i=1}^n q_i a_i^x \right)^{1/x} = \lim_{x \to 0} \frac{\ln \sum_{i=1}^n q_i a_i^x}{x} = \lim_{x \to 0} \frac{\sum_{i=1}^n q_i(a_i^x - 1)}{x}$$

$$= \sum_{i=1}^n q_i \lim_{x \to 0} \frac{a_i^x - 1}{x} = \sum_{i=1}^n q_i \ln a_i. \quad \square$$

Exercise. Deduce from Cauchy's inequality that for all $b_1, b_2, \ldots, b_n \geq 0$, the following inequality holds:

$$\frac{1}{n} \sum_{i=1}^n b_i \geq \left(\frac{1}{n} \sum_{i=1}^n \sqrt{b_i} \right)^2.$$

This yields the monotone decreasing sequence

$$\frac{1}{n} \sum_{i=1}^n a_i \geq \left(\frac{1}{n} \sum_{i=1}^n \sqrt{a_i} \right)^2 \geq \left(\frac{1}{n} \sum_{i=1}^n \sqrt[4]{a_i} \right)^4 \geq \cdots,$$

which by Lemma 1 has limit $\exp\left(\frac{1}{n}\sum_{i=1}^n \ln a_i\right) = \sqrt[n]{\prod_{i=1}^n a_i}$. Hence $\frac{1}{n}\sum_{i=1}^n a_i \geq$
$\sqrt[n]{\prod_{i=1}^n a_i}$. $\square$

This argument has an obvious advantage, and a less obvious disadvantage. The advantage is that it lends itself easily to generalization. Thus, since

$$\sum_{i=1}^n q_i a_i = \sum_{i=1}^n q_i \sum_{i=1}^n q_i a_i \geq \left(\sum_{i=1}^n \sqrt{q_i}\sqrt{q_i a_i}\right)^2 = \left(\sum_{i=1}^n q_i \sqrt{a_i}\right)^2 \geq \cdots,$$

which sequence has (by Lemma 1) limit $\exp(q_i \sum_{i=1}^n \ln a_i) = \prod a_i^{q_i}$, we obtain the following generalization of the a.m.–g.m. inequality:

$$\prod_{i=1}^n a_i^{q_i} \leq \sum_{i=1}^n q_i a_i, \quad \text{where } q_i > 0, \ \sum_{i=1}^n q_i = 1.$$

Exercise. Show that for $q_i \in \mathbb{Q}$, this generalization follows from the original a.m.–g.m. inequality, and then extend to the general situation $q_i \in \mathbb{R}$. (Which property of the exponential function is needed in this step?)

The disadvantage of the fourth proof above has to do with the fact that often in courses in mathematical analysis the a.m.–g.m. inequality is used to establish the basic properties of the exponential function, so that there is some danger of perpetrating a "vicious circle."

Here is the fifth, and last, proof (suffering, incidently, from the same disadvantage). Since the graph of the function $f(x) = \ln x$ is concave down, it follows that if $\sum_{i=1}^n q_i = 1$, $q_i > 0$, then

$$\sum_{i=1}^n q_i f(a_i) \leq f\left(\sum_{i=1}^n q_i a_i\right),$$

that is,

$$\sum_{i=1}^n q_i \ln a_i \leq \ln\left(\sum_{i=1}^n q_i a_i\right), \quad \text{or} \quad \prod_{i=1}^n a_i^{q_i} \leq \sum_{i=1}^n q_i a_i. \quad \square\square$$

Problem 17. Prove that the sequence $\{x_n\}$ where $x_n := (1+1/n)^{n+1}$ is monotone decreasing.

(The desired inequality $x_n < x_{n-1}$ may be written in the form

$$\sqrt[n+1]{\frac{n-1}{n}\cdots\frac{n-1}{n}\cdot 1} < \frac{n}{n+1} = \frac{1}{n+1}\left(n\frac{n-1}{n}+1\right).)$$

Problem 18. Prove the following:
(1) $a_1/a_2 + a_2/a_3 + \cdots + a_n/a_1 \geq n$, $a_i > 0$.
(2) $(1+1/n)^n < 4$.
(3) $\sqrt[n]{n} > \sqrt[n+1]{n+1}$ for $n \geq 3$.

Exercise. From the a.m.–g.m. inequality deduce "Young's inequality":

$$ab \leq \frac{a^p}{p} + \frac{b^q}{q}, \quad a, b, p, q > 0, \quad \frac{1}{p} + \frac{1}{q} = 1.$$

When does equality hold here?

For another proof of Young's inequality, see the solution of Problem 21 below.

4.3   Classical inequalities and geometry

Theorem 3 (Hölder's inequality).   *One has*

$$\sum_{i=1}^{n} a_i b_i \leq \left(\sum_{i=1}^{n} a_i^p \right)^{1/p} \left(\sum_{i=1}^{n} b_i^q \right)^{1/q}, \quad p, q, a_i, b_i > 0, \ 1/p + 1/q = 1.$$

If we write $x_i := a_i^p$, $y_i := b_i^q$, $\alpha := 1/p$, $\beta := 1/q$, the inequality becomes

$$\sum_{i=1}^{n} x_i^\alpha y_i^\beta \leq \left(\sum_{i=1}^{n} x_i \right)^\alpha \left(\sum_{i=1}^{n} y_i \right)^\beta.$$

Dividing throughout by the right–hand side we obtain

$$\sum_{i=1}^{n} \left(\frac{x_i}{\sum_1^n x_i} \right)^\alpha \left(\frac{y_i}{\sum_1^n y_i} \right)^\beta \leq 1.$$

Since $\alpha + \beta = 1$, we have by the generalized a.m.–g.m. inequality that for each i,

$$\left(\frac{x_i}{\sum_1^n x_i} \right)^\alpha \left(\frac{y_i}{\sum_1^n y_i} \right)^\beta \leq \alpha \frac{x_i}{\sum_1^n x_i} + \beta \frac{y_i}{\sum_1^n y_i}.$$

Summation over i now gives the desired inequality.   □

Exercise. Extend Hölder's inequality to the situation where the left–hand side is a sum of products of k factors.

Theorem 4 (Minkowski's inequality).   *One has*

$$\left(\sum_{i=1}^{n} a_i^p \right)^{1/p} + \left(\sum_{i=1}^{n} b_i^p \right)^{1/p} \geq \left(\sum_{i=1}^{n} (a_i + b_i)^p \right)^{1/p}, \quad a_i, b_i > 0, \ p > 1.$$

To see this, set $1/q := 1 - 1/p$, $u_i := a_i + b_i$, $s := \left(\sum_{i=1}^{n} (a_i + b_i)^p \right)^{1/p}$. Then $q > 0$, and by Hölder's inequality,

$$s^p = \sum_{i=1}^{n} u_i^p = \sum_{i=1}^{n} a_i u_i^{p-1} + \sum_{i=1}^{n} b_i u_i^{p-1}$$

$$\leq \left(\sum_{i=1}^{n} a_i^p \right)^{1/p} \left(\sum_{i=1}^{n} u_i^p \right)^{1/q} + \left(\sum_{i=1}^{n} b_i^p \right)^{1/p} \left(\sum_{i=1}^{n} u_i^p \right)^{1/q}.$$

On taking out the factor common to both terms, the last expression becomes

$$s^{p-1}\left(\left(\sum_{i=1}^{n} a_i^p\right)^{1/p} + \left(\sum_{i=1}^{n} b_i^p\right)^{1/p}\right),$$

whence the desired inequality. $\square$

Exercise. Extend Minkowski's inequality to the situation where there are k terms on the left–hand side.

Exercise. Prove that if $p < 1$ then the reverse of Minkowski's inequality holds.

Minkowski's inequality has the following geometric interpretation:

Corollary. *For $p \geq 1$, the map*

$$|\,.\,|_p : \mathbf{x} = (x_1, \ldots, x_n) \in \mathbb{R}^n \longmapsto |\mathbf{x}|_p := \left(\sum_{i=1}^{n} |x_i|^p\right)^{1/p}$$

defines a norm on the vector space $\mathbb{R}^n$

Exercise. Prove that $\lim_{p \to \infty} |\mathbf{x}|_p = \max_{i=1,\ldots,n} |x_i|$.

Prompted by the conclusion of this exercise, we introduce the notation $|\mathbf{x}|_\infty :=$ $\max_{i=1,\ldots,n} |x_i|$. It is clear that $|\,.\,|_\infty$ defines a norm on $\mathbb{R}^n$. Thus do we have exhibited for us a one-parameter family of norms on the space $\mathbb{R}^n$. A natural question arises: What norms can be defined on $\mathbb{R}^n$, and how different are they from one another?

Lemma 2. *Let $p : \mathbb{R}^n \longrightarrow \mathbb{R}$ be a function satisfying the following conditions:*
(1) $p(\mathbf{x}) \geq 0$, and $p(\mathbf{x}) = 0$ precisely if $\mathbf{x} = \mathbf{0}$.
(2) $p(\lambda\mathbf{x}) = |\lambda| p(\mathbf{x})$ for all $\mathbf{x} \in \mathbb{R}^n$, $\lambda \in \mathbb{R}$.
The function p then defines a norm on $\mathbb{R}^n$ if and only if the set $D := \{\mathbf{x} \in \mathbb{R}^n \mid p(\mathbf{x}) \leq 1\}$ is convex.

(The set D is called the *unit ball* with respect to whatever norm is under consideration.)

To prove necessity, observe that if the set D is convex, then for any nonzero vectors $\mathbf{x}, \mathbf{y} \in \mathbb{R}^n$ we have

$$\mathbf{z} := \frac{p(\mathbf{x})}{p(\mathbf{x}) + p(\mathbf{y})} \frac{\mathbf{x}}{p(\mathbf{x})} + \frac{p(\mathbf{y})}{p(\mathbf{x}) + p(\mathbf{y})} \frac{\mathbf{y}}{p(\mathbf{y})} \in D,$$

whence $p(\mathbf{z}) = \dfrac{p(\mathbf{x} + \mathbf{y})}{p(\mathbf{x}) + p(\mathbf{y})} \leq 1$, so that $p(\mathbf{x} + \mathbf{y}) \leq p(\mathbf{x}) + p(\mathbf{y})$, which is the third (and final) defining condition for a norm.

Conversely, if p defines a norm, then for any $\mathbf{x}, \mathbf{y} \in D$ and $\alpha \in [0, 1]$, we have

$$p(\alpha\mathbf{x} + (1 - \alpha)\mathbf{y}) \leq \alpha p(\mathbf{x}) + (1 - \alpha)p(\mathbf{y}) \leq \alpha + (1 - \alpha) = 1,$$

so that $\alpha\mathbf{x} + (1 - \alpha)\mathbf{y} \in D$, i.e., D is convex. $\square$

Theorem 5. *Any two norms on a finite-dimensional real vector space are "equivalent", i.e., given any two such norms p_1, p_2, there exist positive constants c, C such that*

$$cp_1(\mathbf{x}) \leq p_2(\mathbf{x}) \leq Cp_1(\mathbf{x}) \quad \forall \mathbf{x} \in \mathbb{R}^n.$$

Since equivalence of norms on $\mathbb{R}^n$ is easily seen to be an equivalence relation, it suffices to show that any norm p on $\mathbb{R}^n$ is equivalent to the standard Euclidean norm $|\,.\,|_2$. We first show that the function $p : \mathbb{R}^n \longrightarrow \mathbb{R}$ is continuous. To this end let $\{\mathbf{e}_i\}_1^n$ be any basis for $\mathbb{R}^n$, and $\mathbf{x} = \sum_{i=1}^n x_i \mathbf{e}_i$, $\mathbf{y} = \sum_{i=1}^n y_i \mathbf{e}_i$ be any two vectors in $\mathbb{R}^n$, expressed in terms of this basis. Then

$$p(\mathbf{x} - \mathbf{y}) = p\left(\sum_{i=1}^n (x_i - y_i)\mathbf{e}_i\right) \leq \sum_{i=1}^n p(\mathbf{e}_i)|x_i - y_i|$$

$$\leq M \sum_{i=1}^n |x_i - y_i| \leq Mn|\mathbf{x} - \mathbf{y}|_2,$$

where $M := \max_{i=1,\ldots,n}\{p(\mathbf{e}_i)\}$. Hence the function p satisfies the Lipschitz condition, and is therefore certainly continuous.

Now consider the standard unit sphere $S^{n-1} \subset \mathbb{R}^n$. Since S^{n-1} is a compact subset, by Weierstrass's theorem the continuous function p attains a greatest and smallest value on S^{n-1}. Hence there exist positive numbers c and C such that $c \leq p(\mathbf{y}) \leq C$ for all $\mathbf{y} \in S^{n-1}$. For any vector $\mathbf{x} \in \mathbb{R}^n \setminus \{\mathbf{0}\}$, we then have

$$\frac{\mathbf{x}}{|\mathbf{x}|_2} \in S^{n-1} \text{ and } c \leq p\left(\frac{\mathbf{x}}{|\mathbf{x}|_2}\right) = \frac{p(\mathbf{x})}{|\mathbf{x}|_2} \leq C,$$

whence $c|\mathbf{x}|_2 \leq p(\mathbf{x}) \leq C|\mathbf{x}|_2$. $\square$

Corollary. *With respect to every norm on the vector space $\mathbb{R}^n$, the unit ball D is a compact convex subset, with nonempty interior.*

Exercise. Prove that the following two norms on the space $C[0, 1]$ of all functions continuous on $[0, 1]$ are not equivalent:

$$|f|_C := \max_{t \in [0,1]} |f(t)| \text{ and } |f|_1 := \int_0^1 |f|.$$

Theorem 6. *Given any compact, centrally symmetric, convex subset D of $\mathbb{R}^n$, with nonempty interior, there is a norm on $\mathbb{R}^n$ with respect to which this subset is the unit ball.*

Set $p(\mathbf{0}) := 0$, and for $\mathbf{x} \neq \mathbf{0}$, define $p(\mathbf{x}) := \min\{\lambda > 0 \mid \mathbf{x}/\lambda \in D\}$.

Exercise. Prove that this p fulfils the requirements of the theorem. $\square$

Problem 19. Solve the equation $\sin^{19} x + \cos^{92} x = 1$.

(If neither $\sin x$ nor $\cos x$ is zero, then $\sin^{19} x < \sin^2 x$ and $\cos^{92} x < \cos^2 x$, whence $\sin^{19} x + \cos^{92} x < 1$.)

Problem 20. Solve the system

$$\begin{cases} x^3 + y^3 = 1, \\ x^4 + y^4 = 1. \end{cases}$$

The idea behind the solutions of these two problems can be generalized.

Theorem 7 (Jensen's inequality).   *One has*

$$\left(\sum_{i=1}^n a_i^p\right)^{1/p} \geq \left(\sum_{i=1}^n a_i^q\right)^{1/q}, \quad 0 < p \leq q, \ a_i \geq 0,$$

and here if $p < q$, equality occurs precisely if at most one of the numbers a_i, $i = 1, 2, \ldots, n$, is nonzero.

Corollary. *If $1 \leq p \leq q$, then $|\mathbf{x}|_p \geq |\mathbf{x}|_q$ for all $\mathbf{x} \in \mathbb{R}^n$.*

To prove Jensen's inequality, observe first that since both sides of the inequality represent homogeneous functions, it suffices to prove that if $\sum_{i=1}^n a_i^p \leq 1$, then $\sum_{i=1}^n a_i^q \leq 1$. (Why?) Now, if $\sum_{i=1}^n a_i^p \leq 1$, then each $a_i \leq 1$, whence $a_i^q \leq a_i^p$, so $\sum_{i=1}^n a_i^q \leq \sum_{i=1}^n a_i^p \leq 1$. $\square$

In the situation $1 \leq p \leq q$ the omitted step in this proof has the following geometrical interpretation:

Exercise. Let p_1, p_2 be norms on $\mathbb{R}^n$ with respective unit balls D_1, D_2. Prove that

$$D_1 \supset D_2 \iff p_1(\mathbf{x}) \leq p_2(\mathbf{x}) \ \forall \mathbf{x} \in \mathbb{R}^n.$$

4.4   Integral variants of the classical inequalities

We begin this section with two problems whose solutions involve geometric interpretation of the expressions appearing in them.

Problem 21. For any positive real numbers a, b, p, q where $1/p + 1/q = 1$, show that

$$ab \leq \int_0^a x^{p-1} dx + \int_0^b x^{q-1} dx,$$

with equality if and only if $b = a^{p-1}$.

Observe first of all that the condition $1/p + 1/q = 1$ is equivalent to $(p - 1)(q - 1) = 1$, which implies that the functions $f(x) := x^{p-1}$ and $g(x) := x^{q-1}$ are mutual inverses. Hence the integrals on the right-hand side of the inequality in question are equal to the shaded areas in the diagram below, which together contain the rectangle with sides a and b.

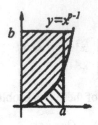

Problem 22. Show that for any continuous and strictly monotonic function f satisfying $f(0) = 0$, $f(1) = 1$, one has

$$f\left(\frac{1}{10}\right) + f\left(\frac{2}{10}\right) + \cdots + f\left(\frac{9}{10}\right) + f^{-1}\left(\frac{1}{10}\right) + \cdots + f^{-1}\left(\frac{9}{10}\right) \leq \frac{99}{10}.$$

Clearly, the quantity

$$\frac{1}{10}\left(f\left(\frac{1}{10}\right) + f\left(\frac{2}{10}\right) + \cdots + f\left(\frac{9}{10}\right)\right)$$

is the sum of the areas of the vertical rectangles shown in the figure below, and similarly, the quantity

$$\frac{1}{10}\left(f^{-1}\left(\frac{1}{10}\right) + \cdots + f^{-1}\left(\frac{9}{10}\right)\right)$$

is the sum of the areas of the horizontal rectangles. It now remains only to observe that all of these rectangles lie in the unit square, that no two of them overlap, and that none of them overlaps with the square of side $1/10$ in the bottom left corner.

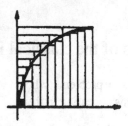

Here is a similar problem.

Problem 23. Prove that

$$9 < \int_0^3 \sqrt[4]{x^4 + 1}\, dx + \int_1^3 \sqrt[4]{x^4 - 1}\, dx < 9.0001.$$

The classical inequalities introduced in the previous sections can be extended to situations where the summations need to be replaced by integrals. Thus let X be a measure space with measure μ, and let $f, g : X \longrightarrow \mathbb{R}$ be measurable functions. Let $p, q > 0$ satisfy $1/p + 1/q = 1$, and suppose that the integrals $A^p := \int_X |f|^p d\mu$ and $B^q := \int_X |g|^q d\mu$ are finite. For each $x \in X$ we have by

Young's inequality

$$\frac{|f(x)g(x)|}{AB} \le \frac{|f(x)|^p}{pA^p} + \frac{|g(x)|^q}{qB^q}.$$

Integrating this inequality over the whole space X, we obtain

$$\frac{1}{AB}\int_X |fg|d\mu \le \frac{1}{pA^p}\int_X |f|^p d\mu + \frac{1}{qA^q}\int_X |g|^q d\mu = \frac{1}{p} + \frac{1}{q} = 1,$$

whence follows the integral version of Hölder's inequality:

$$\int_X |fg|d\mu \le \left(\int_X |f|^p d\mu\right)^{1/p}\left(\int_X |g|^q d\mu\right)^{1/q}.$$

Exercise. Formulate and prove the integral version of Minkowski's inequality.

We conclude this section with the observation that the integral version of Cauchy's inequality (usually called "Schwarz's inequality") has the same geometric meaning as its finite-dimensional analogue:

$$\langle f, g\rangle_{L^2} \le |f|_{L^2}|g|_{L^2}.$$

Provided that f and g are continuous, this follows by essentially the same argument as that establishing Cauchy's inequality.

4.5 Wirtinger's inequality and the isoperimetric problem

In this section we establish the following inequality between the integral of a function and the integral of its derivative.

Theorem 8 (Wirtinger). *Let f be a function of smoothness class C^1 on the interval $[0, 2\pi]$, satisfying $\int_0^{2\pi} f(x)dx = 0$. Then the inequality*

$$\int_0^{2\pi} f^2 \le \int_0^{2\pi} (f')^2$$

holds, with equality if and only if f has the form $f(x) = a\cos x + b\sin x$.

The proof is in essence highly geometrical. If $\{e_k\}_{k=1}^n$ is an orthonormal basis of a finite–dimensional inner–product space, then each vector $\mathbf{x}$ of the space has the resolution $\mathbf{x} = \sum_{k=1}^n \langle \mathbf{x}, e_k\rangle e_k$, where $\langle\ ,\ \rangle$ is the inner product operation. Squaring this equation yields $|\mathbf{x}|^2 = \sum_{k=1}^n \langle \mathbf{x}, e_k\rangle^2$ ("Pythagoras' theorem"). We shall now apply this argument to the space $L^2[0, 2\pi]$ consisting of all square–integrable functions on $[0, 2\pi]$, (i.e., functions f such that the (Lebesgue) integral of f^2 over $[0, 2\pi]$ is finite), furnished with the inner product $\langle f, g\rangle := \int_0^{2\pi} fg$ (guaranteed finite by Schwarz's inequality). It is a routine exercise to show that the functions

$$u_0(x) := \frac{1}{\sqrt{2\pi}}, \quad u_k(x) := \frac{1}{\sqrt{\pi}}\cos kx,$$

$$v_k(x) := \frac{1}{\sqrt{\pi}} \sin kx, \quad k = 1, 2, \ldots,$$

form an orthonormal system (though not a basis!) in this space. Denote by a_k, b_k the Fourier coefficients of an arbitrary function f in the space:

$$a_0 = \frac{1}{\sqrt{2\pi}} \langle f, u_0 \rangle = \frac{1}{2\pi} \int_0^{2\pi} f(x) dx,$$

$$a_k = \frac{1}{\sqrt{\pi}} \langle f, u_k \rangle = \frac{1}{\pi} \int_0^{2\pi} f(x) \cos kx \, dx,$$

$$b_k = \frac{1}{\sqrt{\pi}} \langle f, v_k \rangle = \frac{1}{\pi} \int_0^{2\pi} f(x) \sin kx \, dx.$$

For each natural number n, write

$$f_n := a_0 + \sum_{k=1}^{n} (a_k \cos kx + b_k \sin kx).$$

Then in terms of the above orthonormal system, we have

$$f_n = \langle f, u_0 \rangle u_0 + \sum_{k=1}^{n} (\langle f, u_k \rangle u_k + \langle f, v_k \rangle v_k).$$

Exercise. Show that the difference function $g := f - f_n$ is orthogonal to the vectors $u_0, u_k, v_k, \; k = 1, 2, \ldots$.

It follows that $\langle g, f_n \rangle = 0$, whence

$$|f|_{L^2}^2 = \langle f, f \rangle = \langle g + f_n, g + f_n \rangle = |g|_{L^2}^2 + |f_n|_{L^2}^2.$$

Hence $|f_n|_{L^2}^2 \le |f|_{L^2}^2$; that is,

$$2\pi a_0^2 + \pi \sum_{k=1}^{n} (a_k^2 + b_k^2) \le |f|_{L^2}^2.$$

Since here n is an arbitrary natural number, we infer that

$$2\pi a_0^2 + \pi \sum_{k=1}^{\infty} (a_k^2 + b_k^2) \le |f|_{L^2}^2,$$

which implies, in particular, that the series on the left-hand side converges. Now it turns out that, as is normally shown in first courses in functional analysis, the orthonormal system we are considering here is "complete" (or, equivalently, maximal) with respect to the above-defined inner product, and this is equivalent to the following condition ("Parseval's equation"):

$$2\pi a_0^2 + \pi \sum_{k=1}^{\infty} (a_k^2 + b_k^2) = |f|_{L^2}^2.$$

We are now ready for the proof of Wirtinger's inequality.

Exercise. Prove that if f is continuously differentiable and has Fourier coefficients a_k, b_k, then its derivative f' has Fourier coefficients $a'_k = kb_k$ and $b'_k = -ka_k$.

By assumption $a_0 = \int_0^{2\pi} f = 0$, whence

$$|f|_{L^2}^2 = \pi \sum_{k=1}^{\infty}(a_k^2 + b_k^2) \leq \pi \sum_{k=1}^{\infty}(k^2 b_k^2 + k^2 a_k^2) = |f'|_{L^2}^2,$$

with equality occurring if and only if $a_k = b_k = 0$ for all $k \geq 2$. In this case a further appeal to the completeness of the trigonometric orthonormal system under consideration gives us that in fact $f(x) = a_1 \cos x + b_1 \sin x$. $\square$

Note that it suffices to assume that f is merely piecewise continuously differentiable.

Corollary. If a function f is of class C^1 on the interval $[0, \pi/2]$ and $f(0) = 0$, then $\int_0^{\pi/2} f^2 \leq \int_0^{\pi/2} (f')^2$.

(Consider the function $g(x)$ defined on the interval $[0, 2\pi]$ as follows: For $x \in [0, \pi/2]$, set $g(x) = f(x)$; for $x \in [\pi/2, \pi]$, set $g(x) = f(\pi - x)$; for $x \in [\pi, 3\pi/2]$, set $g(x) = -f(x - \pi)$; and for $x \in [3\pi/2, 2\pi]$, set $g(x) = -f(2\pi - x)$.)

Problem 24. Prove that the integral $-\int_0^{2\pi} f'g$, where $f(t) := a \cos t$, $g(t) := b \sin t$, equals the area enclosed by the ellipse $x^2/a^2 + y^2/b^2 = 1$.

Theorem 9. *Suppose that a plane, smooth, simple, closed curve has length L and encloses a region of area A. Then $L^2 \geq 4\pi A$, with equality precisely when the curve is a circle.*

Let $(x(s), y(s))$, $s \in [0, L]$, be the natural parametrization of the curve (i.e., in terms of (Euclidean) arc length, so that $\dot{x}(s)^2 + \dot{y}(s)^2 = 1$). By choosing the coordinate axes suitably (or, if you like, by translating the curve appropriately), we can arrange that $\int_0^L y(s)ds = 0$. This done, consider the parametrization $(\varphi(t), \psi(t))$ defined in terms of the natural one by

$$\varphi(t) := x\left(\frac{Lt}{2\pi}\right), \quad \psi(t) := y\left(\frac{Lt}{2\pi}\right), \quad t \in [0, 2\pi].$$

By our choice of axes, we have $\int_0^{2\pi} \psi(t)dt = 0$. By Green's theorem, the area of the region bounded by the curve is given by $A = -\int_0^{2\pi} \varphi'\psi$ (assuming the curve traced out so that the enclosed region is on the left). Since

$$(\varphi')^2 + (\psi')^2 = \frac{L^2}{4\pi^2}(\dot{x}^2 + \dot{y}^2) = \frac{L^2}{4\pi^2},$$

we have that

$$\frac{L^2}{2\pi} - 2A = \int_0^{2\pi} ((\varphi')^2 + (\psi')^2) + 2\int_0^{2\pi} \varphi'\psi$$

$$= \int_0^{2\pi} (\varphi' + \psi')^2 + \int_0^{2\pi} ((\psi')^2 - \psi^2) \geq 0,$$

by Wirtinger's inequality. We have equality here precisely if $\psi(t) = a\cos t + b\sin t$ and $\varphi' = -\psi$, i.e., $\varphi(t) = -\int(a\cos t + b\sin t)dt = x_0 + b\cos t - a\sin t$, i.e., if and only if the curve is a circle. $\square$

Corollary (The isoperimetric property of the circle). *Among all piecewise smooth, simple, closed, plane curves of the same length, the circle encloses the region of greatest area.*

In expounding the classical inequalities, the author has followed the book [12], only lightly touching on their connection with the concept of convexity of functions (see in this regard Pólya and Szegő's *Problems and Theorems of Analysis*). In this commentary we shall discuss mainly the inequalities listed in Problem 1, purposely left unsolved (with the one exception).

The effectiveness of any teaching is largely determined by the degree to which the students are involved in the process. As everyone knows well who has had any experience of teaching to mathematical circles, it is useful to set problems in series, and the best textbooks devoted to providing problems suited to such groups do organize the problems serially. Of course, any list of problems might be termed a "series," and in the classroom mathematical problems are not normally suggested singly. The following problem–series are the most common: The first is that of the "special case": having described a method for solving, for instance, quadratic inequalities, the teacher then provides a collection of exercises to be solved using that method. The second is the "single–concept" type, where the solutions of the problems of the series all involve variations of the same idea. For instance, all of the problems in Section 1 of the present chapter may together be considered as comprising a series on the theme of the inequality $a^2 + b^2 \geq 2|ab|$ (although of course varying in technical complexity). The third type is that of the "step-by-step" approach, where the teacher formulates a sequence of statements culminating in a nonobvious assertion; the students are thus shown the route, as it were, that they should follow to obtain in the end an interesting result. An example of this type of problem–series is furnished by inequality (5) of Problem 1 followed by parts (a), (b), and (c) of Problem 9.

The next, and last, type is illustrated by the inequalities (13), (14), (12) of Problem 1 (in that order!). The first of these is very simple. That one out of the way, the students have then to see (?!) that since $(a+b-c)+(b+c-a) = 2b$, the substitution $u = (b+c-a)/2$, $v = (c+a-b)/2$, $w = (a+b-c)/2$ in the second inequality yields the first! And this same substitution is the key to solving the third inequality. (Try to solve it without using that substitution.) The author's name for this type of problem–series is "trail–of–clues."

CHAPTER 5

Sets, Equations, and Polynomials

5.1 Figures and their equations

Problem 1. Sketch the plane sets defined by (i.e., the graphs of) each of the following equations and inequalities: $x^2 = 1$; $x^2 + y^2 = 2y - 1$; $x^2 + y^2 = 2xy$; $xy + 1 = x + y$; $x^2 + y^2 = 2x$; $x^2 y + y^3 = 2xy$; $|x - y| = 1$; $x^2 + xy - 2y^2 = 0$; $y^2 \le |x|$; $\max\{|x|, |y|\} \le 1$; $|x + y| + |x - y| \le 2$.

Some of the answers are shown in the following diagrams.

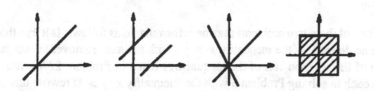

In the course of solving this problem one repeatedly needs to use particular instances of the general facts formulated in the following theorem, which might be called—with some exaggeration perhaps, although not altogether inappropriately—the "fundamental theorem of analytic geometry."

Theorem 1. *Let M and N be the respective solution sets of the equations $f(x, y) = 0$ and $g(x, y) = 0$. Then:*

(1) *The solution set of the equation $f(x, y)g(x, y) = 0$ is the union $M \cup N$.*

(2) *the equation $f^2(x, y) + g^2(x, y) = 0$ and the system $\begin{cases} f(x, y) = 0, \\ g(x, y) = 0 \end{cases}$ each have as solution set the intersection $M \cap N$.*

Recall the basic definition: We say that a subset $\mathcal{M}$ of the plane (furnished with some coordinate system—below it will be a rectangular Cartesian coordinate system) is *defined by* (or is the *solution set*, or *graph*, of) an equation $f(x, y) = 0$ (or an inequality $f(x, y) > 0$), under the following two conditions:

(i) the coordinates of every point in the given set $\mathcal{M}$ satisfy the given equation (or inequality);

(ii) for every ordered pair (x, y) satisfying the given equation (or inequality), the point P with coordinates (x, y) belongs to $\mathcal{M}$.

Here is the proof of part (1) of the theorem. If $P(x_0, y_0)$ is a point in $\mathcal{M} \cup \mathcal{N}$, then either $P \in \mathcal{M}$ or $P \in \mathcal{N}$, whence by part (i) of the definition, either $f(x_0, y_0) = 0$ or $g(x_0, y_0) = 0$. In any case, $f(x_0, y_0)g(x_0, y_0) = 0$, so that the coordinates (x_0, y_0) of P satisfy the equation $f(x, y)g(x, y) = 0$. Conversely, if $f(x_0, y_0)g(x_0, y_0) = 0$, then either $f(x_0, y_0) = 0$ or $g(x_0, y_0) = 0$, whence by condition (ii) of the definition, we have either $P(x_0, y_0) \in \mathcal{M}$ or $P(x_0, y_0) \in \mathcal{N}$, i.e., $P \in \mathcal{M} \cup \mathcal{N}$. $\square$

Exercise. Formulate and prove the analogous result for the set defined by the inequality $f(x, y)g(x, y) > 0$.

Exercise. Find an algebraic proof of the set-theoretic formula $\mathcal{A} \cup (\mathcal{B} \cap \mathcal{C}) = (\mathcal{A} \cup \mathcal{B}) \cap (\mathcal{A} \cup \mathcal{C})$.

The following two problems appear together not just by chance.

Problem 2. Is it true that if $a > b$, $a, b \neq 0$, then: $1/a < 1/b$; $a^2 > b^2$; $a^3 > b^3$; $a/b > 1$; $b/a < 1$; $a^3 - 3a > b^3 - 3b$?

Problem 3. Sketch in the coordinate plane the solution sets of the following inequalities: $1/x < 1/y$; $x^2 > y^2$; $x^3 > y^3$; $x/y > 1$; $y/x < 1$; $x^3 - 3x > y^3 - 3y$?

The first of these two problems may be reformulated as follows: Is it true that the half-plane defined by the inequality $x > y$ (with the axes removed) is contained in each of the solution sets of the inequalities listed in Problem 2? We illustrate an approach to solving Problem 3 with the inequality $x/y > 1$: rewrite this in the equivalent form $(x - y)/y > 0$, and apply to this inequality the answer to the first exercise above.

Thus it is for the most part not difficult to see that in particular, the solution sets of the inequalities in Problem 3 are all different. In using the phrase "for the most part" here, we have in mind the last inequality in that problem, which may be rewritten in the form $(x - y)(x^2 + xy + y^2 - 3) > 0$. For now we shall postpone analyzing this inequality (see Problem 5). Here instead is a problem similar to Problem 1.

Problem 4. Sketch in the coordinate plane the sets defined by the following inequalities:

$\{x\} \leq \{y\}$, where $\{\,.\,\}$ indicates the "fractional part" of a real number.

$\sin(x + y) \sin(x - y) \geq 0$.

$$|x| \le |\sin y|.$$
$$(x^2 + y^2 - 4)\sqrt{1 - |x|} \le 0.$$

Problem 5. Prove that the set defined by the inequality $x^2 + xy + y^2 \le 3$:

(a) is symmetric with respect to the origin;

(b) contains the disk of radius $\sqrt{2}$, center the origin;

(c) is contained in the disk of radius $\sqrt{6}$, center the origin;

(d) is contained in the square of side 4 with center the origin.

For (a), observe simply that any two pairs (x, y), $(-x, -y)$ together either both satisfy the given inequality or do not. Hence any two points $P(x, y)$, $Q(-x, -y)$ that are mutual images under reflection in the origin either both belong to the solution set or both lie outside it, so that indeed that set is invariant under that reflection.

For (b) and (c) we argue as follows: From the familiar inequality $2|xy| \le x^2 + y^2$, we deduce that $(x^2 + y^2)/2 \le x^2 + xy + y^2 \le 3(x^2 + y^2)/2$. Hence if $x^2 + y^2 \le 2$, then $x^2 + xy + y^2 \le 3$, and if $x^2 + xy + y^2 \le 3$, then $x^2 + y^2 \le 6$.

(d) Since $x^2 + xy + y^2 = 3x^2/4 + (x/2 + y)^2$, it follows that if $x^2 + xy + y^2 \le 3$, then we must certainly have $3x^2/4 \le 3$, i.e., $|x| \le 2$. Then by symmetry (a different one!) we obtain also $|y| \le 2$.

Those familiar with the classification of quadratic curves will doubtless have realized that the solution set of the inequality of Problem 5 is the region bounded by the ellipse defined by the corresponding equation $x^2 + xy + y^2 = 3$. Here the power of the method of coordinates is demonstrated by our ability to elicit various properties of this figure by simple algebraic means.

Problem 6. Let $\mathcal{M}$ be the graph of the equation $f(x, y) = 0$. Write down the equation defining the set obtained from $\mathcal{M}$ by (a) translating it through the vector $v(k, l)$; (b) reflecting it in the point $P(x_0, y_0)$; (c) applying to it the dilation by the factor k with center the origin.

In order to avoid errors in solving this and similar problems, one should give careful step-by-step arguments from the definitions. Consider, for instance, part (c) of the above problem: A point $P(x, y)$ belongs to the image set of $\mathcal{M}$ under the dilation if and only if the point P' with coordinates (x', y') satisfying $x = kx'$, $y = ky'$ belongs to $\mathcal{M}$, i.e., if and only if the point $P(x/k, y/k)$ lies in $\mathcal{M}$, i.e., if and only if $f(x/k, y/k) = 0$. This then is the equation defining the image of the set $\mathcal{M}$.

We conclude this introductory section with the following two–variable version of the "interval method," as it is sometimes called, for solving inequalities in one variable. For simplicity, we confine ourselves to the situation where the functions appearing in the inequality are defined on the whole plane.

Theorem 2. *The solution set of the inequality $f(x, y)g(x, y) > 0$ (or, equivalently,*
$$\frac{f(x, y)}{g(x, y)} > 0), \text{ where } f, g : \mathbb{R}^2 \longrightarrow \mathbb{R} \text{ are continuous functions of } x \text{ and } y, \text{ is the}$$
union of connected components of the complement in $\mathbb{R}^2$ of the subset Z defined by

the equation $f(x, y)g(x, y) = 0$. *(To put it more simply, the set Z divides the plane up into pairwise disjoint parts, each of which either consists entirely of solutions of the given inequality or contains no solutions.)*

The idea of "connectedness" of a set (and of the "connected components" of a set) is one of the basic concepts of topology, that subdomain of mathematics devoted to the study of the idea of continuity in its most general sense. In our present context, namely the plane $\mathbb{R}^2$, we may take the following as the definition of *connected component*: Two points P and Q belong to the same connected component of the set $\mathbb{R}^2 \setminus Z$ if they can be joined by a (continuous) polygonal arc (broken line segment) lying entirely in that set, i.e., not intersecting Z.

Here is the proof of the theorem. Suppose that the point $P(x_0, y_0)$ represents a solution of the inequality in question, and that $Q(x', y')$ is any point in the same connected component of $\mathbb{R}^2 \setminus Z$ as P.

Exercise. Let $P = A_0(x_0, y_0), A_1(x_1, y_1), \ldots, A_n(x', y') = Q$ be the vertices of a polygonal arc joining P and Q (and lying in $\mathbb{R}^2 \setminus \mathcal{M}$). Find formulae for functions $\varphi, \psi : [0, 1] \longrightarrow \mathbb{R}^2$ satisfying $\varphi(0) = x_0, \varphi(1) = x', \psi(0) = y_0, \psi(1) = y'$, such that the map $\mathbf{r} : t \mapsto (\varphi(t), \psi(t))$ affords a continuous parametrization of this polygonal arc.

(The functions φ and ψ may be chosen piecewise linear (piecewise "affine," more precisely).)

Now consider the function $u : [0, 1] \longrightarrow \mathbb{R}^2$ defined by the formula

$$u(t) := f(\varphi(t), \psi(t))\, g(\varphi(t), \psi(t)).$$

Since $u(0) = f(x_0, y_0)\, g(x_0, y_0) > 0$ and since u is continuous and nowhere vanishing on $[0, 1]$, it follows from the intermediate–value theorem that $u(t) > 0$ for all $t \in [0, 1]$. In particular, $u(1) = f(x', y')\, g(x', y') > 0$, i.e., the point (x', y') is a solution of the given inequality. $\square$

5.2   Pythagorean triples and Fermat's last theorem

Problem 7. Show that the formulae $x = 2kmn$, $y = k(m^2 - n^2)$, $z = k(m^2 + n^2)$, where k, m, n are integers, give all (up to interchanging x and y) integer solutions of the equation $x^2 + y^2 = z^2$.

That these formulae do give solutions is easily checked by simply substituting them in the equation.

Lemma 1. *The formulae $x = 2t/(t^2 + 1)$, $y = (t^2 - 1)/(t^2 + 1)$, where $t \in \mathbb{Q}$, give all rational solutions of the equation $x^2 + y^2 = 1$, except $x = 0$, $y = 1$.* $\square$

Consider the family of straight lines through the point $S(0, -1)$, parametrized by their slope t. It is straightforward to verify that the point of intersection P of the unit circle with the line through S of slope t has coordinates $(2t/(t^2 + 1), (t^2 -$

$1)/(t^2 + 1))$. The lemma now follows by observing that each rational value of t yields a rational point P, i.e., a point with rational coordinates, and conversely, if the point P is rational, then the slope t of the line SP is also rational.

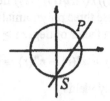

Returning to the solution of Problem 7, we denote by a, b, c any solution in integers of the equation $x^2 + y^2 = z^2$. Since the trivial solution $0, 0, 0$ clearly fits the description, we may assume that $c \neq 0$, and then, by interchanging the roles of a and b if necessary, that also $a \neq 0$. Then by Lemma 1 there is a nonzero rational number $t = p/q$, $p, q \in \mathbb{Z}$, such that

$$\frac{a}{c} = \frac{2t}{t^2 + 1} = \frac{2pq}{p^2 + q^2} \quad \text{and} \quad \frac{b}{c} = \frac{t^2 - 1}{t^2 + 1} = \frac{p^2 - q^2}{p^2 + q^2}.$$

Of course we may, and shall, assume that p and q are relatively prime.

Exercise. Prove that if p and q are relatively prime and one of them is even, then the integers $2pq$ and $p^2 + q^2$ are likewise relatively prime.

Hence if one of p, q is even, then the fraction $2pq/(p^2 + q^2)$ is in its lowest terms, so that $a = 2kpq, c = k(p^2 + q^2)$, whence $b = k(p^2 - q^2)$. On the other hand, if both p and q are odd, say $p = 2u + 1, q = 2v + 1$, then

$$2pq = 2((u + v + 1)^2 - (u - v)^2),$$
$$p^2 - q^2 = 4(u - v)(u + v + 1),$$
$$p^2 + q^2 = 2((u + v + 1)^2 + (u - v)^2).$$

Hence setting $m := u + v + 1, n := u - v$, we have $a/c = (m^2 - n^2)/(m^2 + n^2)$, $b/c = 2mn/(m^2 + n^2)$. If m and n are both odd, then one repeats the last step; since $m^2 + n^2 < u^2 + v^2$, we have a descent that must halt after a finite number of steps.

The elegance of Lemma 1, giving a formula for all rational points on the unit circle, might tempt one into seeking an analogous formula for the rational solutions of the equation $x^n + y^n = 1$ for $n \geq 3$, which would bring one up against Fermat's last theorem: For $n \geq 3$ the curve $x^n + y^n = 1$ has no rational points. Of course we are not going to prove this theorem here; instead, we shall have to rest content with the following very much easier result [28].

Theorem 3. *For $n \geq 3$, the curve $x^n + y^n = 1$ has no rational parametrization, i.e., cannot be parametrized in terms of rational functions of a parameter t with real coefficients. (In fact somewhat more is true: there are no nonconstant rational* *functions $p(t)/r(t)$, $q(t)/r(t)$ over $\mathbb{R}$ the sum of whose nth powers is identically equal to 1.)*

Suppose that there exist real polynomials $p(t)$, $q(t)$ and $r(t)$, not all constant, such that for all t the point $(p(t)/r(t), q(t)/r(t))$ lies on the curve $x^n + y^n = 1$. We may clearly suppose that these three polynomials have highest common factor 1, i.e. that they have no common factor of degree ≥ 1. Thus

$$p^n(t) + q^n(t) - r^n(t) \equiv 0,$$

and differentiation of this identity yields

$$p'(t)p^{n-1}(t) + q'(t)q^{n-1}(t) - r'(t)r^{n-1}(t) \equiv 0.$$

From these two identities we see that $x = p^{n-1}$, $y = q^{n-1}$, $z = r^{n-1}$ is a solution of the system

$$px + qy - rz = 0,$$
$$p'x + q'y - r'z = 0.$$

Since every solution of this system is proportional to the (nonzero) triple $(q'r - qr', r'p - rp', p'q - pq')$, there exist relatively prime polynomials f and g such that

$$p^{n-1} = \frac{f}{g}(q'r - qr'), \quad q^{n-1} = \frac{f}{g}(r'p - rp'), \quad r^{n-1} = \frac{f}{g}(p'q - pq').$$

The relative primeness of p, q, and r implies that of p^{n-1}, q^{n-1}, and r^{n-1}, whence it follows that $f(t) \equiv 1$. Let k, l, m be the degrees of p, q, r respectively, and suppose, for instance, that $k \geq l \geq m$. Since $q'r - qr' = gp^{n-1}$, the degree of the polynomial p^{n-1} is at most that of $q'r - qr'$, which in turn is at most $l + m - 1$. Hence $(n - 1)k \leq l + m - 1$, which is false if $n \geq 3$. Similar contradictions are obtained in the other cases (or the symmetry of the above equations may be invoked). $\square$

Parametric representation of a set may arise naturally also in solving elementary problems.

Problem 8. Sketch in the coordinate plane the set of all points (a, b) such that the equation $x^3 + ax + b = 0$ has at most two distinct solutions.

Suppose that the polynomial $x^3 + ax + b$ has as its roots t, t, u (i.e., t is a repeated root). Then by the factor theorem (sometimes attributed to Bézout) we have $x^3 + ax + b = (x - t)^2(x - u)$, whence $2t + u = 0$, $t^2 + 2ut = a$, and $b = -ut^2$, yielding $a = -3t^2$ and $b = 2t^3$. Hence the desired set consists of the points of the form $(-3t^2, 2t^3)$, $t \in \mathbb{R}$. To find the nonparametric equation of this curve, we eliminate t between $x = -3t^2$ and $y = 2t^3$ as follows. The equation $x = -3t^2$ is equivalent to $x^3 = -27t^6$, and replacing t^3 by $y/2$ in the latter equation, we get $x^3 = -27y^2/4$ or $4x^3 + 27y^2 = 0$ as the equation of the curve (sketched in the diagram).

Exercise. Prove that the number of real roots of the equation $x^3 + ax + b = 0$ (one or three) is determined by the sign of the expression $-4a^3 - 27b^2$.

(Here the simplest approach is to investigate the behavior of the function $y = x^3 + ax$. Try also to find a solution using purely algebraic means.)

Lemma 2. *The substitution given by $\sqrt{ax^2 + bx + c} = tx \pm \sqrt{c}$ yields a rational parametrization of part of the curve $y^2 = ax^2 + bx + c$, $c > 0$.*

The straight line given by the equation $y = tx + \sqrt{c}$ meets the given curve in two points. The obvious one of these is $(0, \sqrt{c})$, and the other we find by solving $(tx + \sqrt{c})^2 = ax^2 + bx + c$ or $t^2x^2 + 2t\sqrt{c}x = ax^2 + bx$, yielding $x = 0$, already noted, and $x = 2t\sqrt{c}b/(a - t^2)$, with corresponding y–coordinate $y = ((2 - \sqrt{c})t^2 + a\sqrt{c})/(a - t^2)$. $\square$

Exercise. What is the geometric idea behind the substitution defined by $\sqrt{ax^2 + by + c} = \pm\sqrt{a}\,x + t$, $a > 0$?

Corollary. *For any polynomial $P(x, y)$, the integral (antiderivative) $\int P(x, \sqrt{ax^2 + bx + c})\,dx$ can be evaluated explicitly in terms of elementary functions.*

To see this, consider any rational parametrization $x = \varphi(t)$, $y = \psi(t)$ of the curve $y^2 = ax^2 + bx + c$. (That there is such a parametrization is guaranteed by Lemma 2.) The substitution $x = \varphi(t)$ converts the given integral into the form $\int P(\varphi(t), \psi(t))\varphi'(t)dt$, an integral of a rational function. $\square$

Problem 9. Find a rational parametrization of the curve $x^3 + y^3 = 3xy$, and sketch this curve.

We conclude this section with an example and an argument.

It is not difficult to see that the equation

$$x^4 + 2x^2y^2 + y^4 - 3x^2 - 3y^2 + 2 = 0$$

has as its graph two circles, with one inside the other. Consider the general quartic equation $p(x, y) = 0$, i.e., with p a polynomial of degree 4. Impose the condition that the curve defined by this equation include two closed curves (called "ovals") with one inside the other. We shall show that these two ovals must then in fact constitute the whole curve. Suppose on the contrary that there is a point on neither oval whose coordinates satisfy the given equation $p(x, y) = 0$. Let $ax + by + c = 0$ be the equation of the straight line passing through this point and some other point lying inside the smaller of the two ovals (see the diagram). Clearly, this straight line meets the graph of the equation $p(x, y) = 0$ in at least five distinct points. However, these five points correspond to solutions of the system

$$p(x, y) = 0,$$
$$ax + by + c = 0,$$

and on substituting for y in $p(x, y) = 0$ using $y = -(ax + c)/b$ (if $b \neq 0$), we obtain an equation in x of degree at most four, and so with not more than four solutions.

5.3   Numbers and polynomials

Problem 10. Simplify the expression

$$\frac{(x - a)(x - b)}{(c - a)(c - b)} + \frac{(x - b)(x - c)}{(a - b)(a - c)} + \frac{(x - c)(x - a)}{(b - c)(b - a)},$$

where a, b, c are distinct.

Denote this quadratic expression by $P(x)$. It is easy to check that $P(a) = P(b) = P(c) = 1$. Thus the equation $P(x) = 1$ has three distinct solutions, which is possible only if the polynomial P is identically equal to 1 (why?).

In this solution we have used a particular case of the following result,

Theorem 4. *If a polynomial* $p(x) = a_n x^n + a_{n-1} x^{n-1} + \cdots + a_0$ *of degree at most* n *has more than* n *roots, then its coefficients are all zero.*

The usual proof of this theorem uses the factor theorem. Here is an alternative argument: Suppose that $x_0, x_1, \ldots, x_n$ are distinct roots of the polynomial p. Then

$$a_0 x_0^n + a_1 x_0^{n-1} + \cdots + a_n = 0,$$
$$a_0 x_1^n + a_1 x_1^{n-1} + \cdots + a_n = 0,$$
$$\cdots\cdots\cdots\cdots\cdots\cdots\cdots\cdots\cdots$$
$$a_0 x_n^n + a_1 x_n^{n-1} + \cdots + a_n = 0.$$

Hence if any of the exhibited coefficients of the polynomial p were nonzero, then the homogeneous linear system with matrix $X = \left(x_i^{n-j} \right)_{i,j=0}^{n}$ would have a nontrivial solution. This, however, is not possible, since the determinant of this matrix—the "Vandermonde determinant"—is nonzero:

$$\det X = \prod_{0 \le k < l \le n} (x_k - x_l) \ne 0. \quad \square$$

Exercise. Show that two different polynomials in two (or more) variables can take the same values at infinitely many points.

Corollary. *If f and g are polynomials in n variables which coincide at every point $(x_1, x_2, \ldots, x_n) \in \mathbb{R}^n$, then $f = g$, (i.e., they have the same coefficients).*

Exercise. Prove this corollary.

Compare the two solutions given for the following problem.

Problem 11. Prove that if $abc = 1$ and $1/a + 1/b + 1/c = a + b + c$, then at least one of a, b, c is equal to 1.

First solution: Since

$$(a - 1)(b - 1)(c - 1) = abc - 1 + ab + bc + ca - (a + b + c) =$$
$$abc \left(\frac{1}{a} + \frac{1}{b} + \frac{1}{c} \right) - (a + b + c) = \frac{1}{a} + \frac{1}{b} + \frac{1}{c} - (a + b + c) = 0,$$

we must have either $a = 1$ or $b = 1$ or $c = 1$.

Second solution: Consider the polynomial

$$P(x) := (x - a)(x - b)(x - c),$$

with roots the given numbers. Since $abc = 1$ and $a + b + c = ab + bc + ca$, we have

$$P(x) = x^3 - Ax^2 + Ax - 1 = (x - 1)(x^2 + (1 - A)x + 1),$$

which has 1 as a root.

Problem 12. Show that (a) the polynomial $x^3 - 7x^2 + 5x - 11$ has no negative roots; (b) if a, b, c are real numbers such that abc, $ab + bc + ca$ and $a + b + c$ are all positive, then a, b and c are themselves positive.

Part (a) is proved by contradiction. For part (b) consider the polynomial P of the second solution of the preceding problem, applying to it the argument proving part (a).

5.4   Symmetric polynomials

Problem 13. Find the sum of the cubes of the roots of each of the following two equations:
(1) $x^2 - 5x - 7 = 0$.
(2) $2x^3 - 3x^2 - 12x - 1 = 0$.

For (1) one may use an identity:

$$x_1^3 + x_2^3 \equiv (x_1 + x_2)\left((x_1 + x_2)^2 - 3x_1x_2\right) = 5(5^2 + 21) = 230,$$

or instead proceed more directly:

$$x_1^3 + x_2^3 = 5(x_1^2 + x_2^2) + 7(x_1 + x_2)$$
$$= 5(5(x_1 + x_2) + 14) + 7(x_1 + x_2) = 32(x_1 + x_2) + 70 = 230,$$

and similarly for equation (2):

$$x_1^3 + x_2^3 + x_3^3 = \frac{1}{2}\left(3(x_1^2 + x_2^2 + x_3^2) + 12(x_1 + x_2 + x_3) + 3\right)$$
$$= \frac{1}{2}\left(3(x_1 + x_2 + x_3)^2 - 6(x_1x_2 + x_2x_3 + x_3x_1)\right.$$
$$\left. + 12(x_1 + x_2 + x_3) + 3\right) = \cdots,$$

where in the final (omitted) step one again applies Viète's theorem (expressing the coefficients of a polynomial in terms of its roots).

The first method used in connection with equation (1) can also be used for (2), although the analogous identity is harder to come by:

$$x^3 + y^3 + z^3 = 3xyz + (x + y + z)\left((x + y + z)^2 - 3(xy + yz + zx)\right)$$
$$= (x + y + z)^3 - 3(x + y + z)(xy + yz + zx) + 3xyz.$$

Exercise. Show that

$$x^4 + y^4 + z^4 = (x + y + z)^4 - 4(x + y + z)^2(xy + yz + zx)$$
$$+ 2(xy + yz + zx)^2 + 4xyz(x + y + z).$$

Problem 14. Prove that if a, b, c are positive real numbers satisfying $2(a^8 + b^8 + c^8) = (a^4 + b^4 + c^4)^2$, then the triangle with sides a, b, c is right–angled.

(Use $(x^2 + y^2 + z^2)^2 - 2(x^4 + y^4 + z^4) = (x + y + z)(x + y - z) \times (y + z - x)(z + x - y)$.)

We now define the "elementary symmetric polynomials." Consider the following polynomial in the $n + 1$ variables $x, x_1, x_2, \ldots, x_n$: $S(x) := \prod_{i=1}^{n}(x - x_i)$. Expanding this in powers of x, we obtain

$$S(x) = \prod_{i=1}^{n}(x - x_i) = \sum_{k=0}^{n}(-1)^k x^{n-k} s_k^{(n)}(x_1, \ldots, x_n).$$

The *elementary symmetric polynomials in n variables* are then just the polynomials $s_k^{(n)}(x_1, \ldots, x_n)$. (It is easy to see that $s_k^{(n)}$ is the sum of all products of the form $x_{i_1} x_{i_2} \ldots x_{i_k}$, $1 \le i_1 < \cdots < i_k \le n$.)

More generally, a polynomial $f(x_1, \ldots, x_n)$ is said to be *symmetric* if

$$f(x_1, \ldots, x_n) = f(x_{i_1}, \ldots, x_{i_n})$$

for every permutation $(i_1, \ldots, i_n)$ of the set $\{1, 2, \ldots, n\}$.

Exercise. Show that the elementary symmetric polynomials $s_k^{(n)}$ are indeed symmetric.

It is obvious that for any polynomial $F(y_1, \ldots, y_r)$, the polynomial in $x_1, \ldots, x_n$ given by

$$F\left(s_{j_1}^{(n)}(x_1, \ldots, x_n), \ldots, s_{j_r}^{(n)}(x_1, \ldots, x_n)\right),$$

where $1 \le j_1, \ldots, j_r \le n$, is symmetric. It is interesting (and nontrivial) that every symmetric polynomial can be obtained in this way [19].

Theorem 5. *Given a symmetric polynomial* $f(x_1, \ldots, x_n)$, *there is a unique polynomial* $F(y_1, \ldots, y_n)$ *such that*

$$f(x_1, \ldots, x_n) = F\left(s_1^{(n)}(x_1, \ldots, x_n), \ldots, s_n^{(n)}(x_1, \ldots, x_n)\right).$$

(In other words, the ring consisting of all symmetric polynomials is a ring of polynomials in the elementary symmetric polynomials.)

We shall prove the existence of such a polynomial F by means of a double induction on the number of variables n and the degree m of the given symmetric polynomial f. If $m = 0$ (or if $f \equiv 0$), take $F := f$ ($\equiv$ const.). Note that if $n = 0$, then $m = 0$ (or $f \equiv 0$). Let $m > 0$, $n > 0$, and suppose inductively that such a polynomial F exists for every symmetric polynomial f in at most $n - 1$ variables or of degree at most $m - 1$.

Consider an arbitrary symmetric polynomial $f(x_1, \ldots, x_n)$ of degree m. By the inductive assumption there is a polynomial $g_1(y_1, \ldots, y_{n-1})$ such that

$$f(x_1, \ldots, x_{n-1}, 0) = g_1\left(s_1^{(n-1)}(x_1, \ldots, x_{n-1}), \ldots, s_{n-1}^{(n-1)}(x_1, \ldots, x_{n-1})\right).$$

Now consider the difference

$$f_1(x_1, \ldots, x_n) := f(x_1, \ldots, x_n) - g_1\left(s_1^{(n)}(x_1, \ldots, x_n), \ldots, s_{n-1}^{(n)}(x_1, \ldots, x_n)\right).$$

Since $f_1(x_1, \ldots, x_{n-1}, 0) = 0$, the polynomial f_1 must be divisible by x_n, and therefore, in view of its symmetry,

$$\begin{aligned} f_1(x_1, \ldots, x_n) &= x_1 \cdots x_n\, h(x_1, \ldots, x_n) \\ &= s_n^{(n)}(x_1, \ldots, x_n)\, h(x_1, \ldots, x_n), \end{aligned}$$

where h is also symmetric. Since $\deg h < m$, by the inductive assumption the polynomial h, and hence also f, is representable as a polynomial in $s_1^{(n)}, \ldots, s_n^{(n)}$.

For the uniqueness of F, it suffices to establish the algebraic independence of the elementary symmetric polynomials, i.e., to show that if

$$g\left(s_1^{(n)}(x_1, \ldots, x_n), \ldots, s_n^{(n)}(x_1, \ldots, x_n)\right)$$

is identically zero as a polynomial in $x_1, \ldots, x_n$, then $g(y_1, \ldots, y_n)$ must likewise be identically zero.

Assume inductively that this holds with n replaced throughout by $n - 1$, and let $g(y_1, \ldots, y_n)$ be a polynomial of least degree such that

$$g\left(s_1^{(n)}(x_1, \ldots, x_n), \ldots, s_n^{(n)}(x_1, \ldots, x_n)\right) \equiv 0.$$

Write this polynomial in the form

$$g(y_1, \ldots, y_n) = g_0(y_1, \ldots, y_{n-1}) + \sum_{k=1}^{d} g_k(y_1, \ldots, y_{n-1}) y_n^k.$$

Exercise. Prove that the polynomial g_0 is nonzero.

Now, setting $x_n = 0$ in the identity $g(s_1^{(n)}, \ldots, s_n^{(n)}) \equiv 0$, we obtain $g_0(s_1^{(n-1)}, \ldots, s_{n-1}^{(n-1)}) \equiv 0$, whence by the inductive assumption, $g_0(y_1, \ldots, y_{n-1}) \equiv 0$. $\square$

Corollary. *If $f(x_1, \ldots, x_n)$ is a symmetric polynomial, and $t_1, \ldots, t_n$ are the roots of the equation $x^n + \sum_{k=1}^{n} a_k x^{n-k} = 0$, then there exists a polynomial $F(y_1, \ldots, y_n)$ such that*

$$f(t_1, \ldots, t_n) = F(a_1, \ldots, a_n).$$

Moreover, if the coefficients of f are from a (commutative) ring R, then the coefficients of F also come from R.

Problem 15. Prove that if $x + x^{-1}$ is an integer, then $x^k + x^{-k}$ is an integer for all $k \in \mathbb{Z}$.

(Use the identity $(a^{k-1} + b^{k-1})(a + b) = a^k + b^k + ab(a^{k-2} + b^{k-2})$.)

The idea behind the solution of this problem can be used to express the polynomials $p_k^{(n)} = \sum_{i=1}^{n} x_i^k$ as polynomials in the elementary symmetric functions.

Problem 16. Prove that if a is an odd integer, and x, y are the roots of the polynomial $t^2 + at - 1$, then $x^4 + y^4$ and $x^5 + y^5$ are relatively prime integers.

Problem 17. Prove that the first integer greater than $(3 + \sqrt{5})^n$ is divisible by 2^n.

(The number $(\frac{3+\sqrt{5}}{2})^n + (\frac{3-\sqrt{5}}{2})^n$ is an integer.)

5.5 Discriminants and resultants

Let us apply the theorem of the preceding section—or rather its corollary—to the symmetric polynomial

$$\Delta_n^2 = \left(\det(x_i^j)_{i=1,\ldots,n}^{j=0,\ldots,n-1}\right)^2,$$

the square of the Vandermonde determinant. If we regard $x_1, \ldots, x_n$ as the roots of the general nth degree monic polynomial $f(x) = x^n + a_1 x^{n-1} + \cdots + a_n$, then

$a_k = (-1)^k s_k^{(n)}$, $k = 1, \ldots, n$, so that by Theorem 5 there is a unique polynomial $D_n(y_1, \ldots, y_n)$ such that $D_n(a_1, \ldots, a_n) = \Delta_n^2$. The polynomial D_n is called the *discriminant* of the (general) polynomial f.

Theorem 6. *A polynomial* $f(x) = x^n + a_1 x^{n-1} + \cdots + a_n$ *has multiple roots (in some appropriate field, e.g. the field of complex numbers if the coefficients are numbers) if and only if* $D_n(a_1, \ldots, a_n) = 0$.

This is almost immediate, since if $t_1, \ldots, t_n$ are the roots of f, then

$$D_n(a_1, \ldots, a_n) = \Delta_n^2(t_1, \ldots, t_n) = \prod_{1 \le i < j \le n} (t_i - t_j)^2,$$

and this product is zero if and only if at least two of the roots coincide. $\square$

In solving Problem 8 we showed that the polynomial $x^3 + ax + b$ has repeated roots if and only if $4a^3 + 27b^2 = 0$. We shall now show that in fact $D(a, b) = -4a^3 - 27b^2$ is the discriminant of this polynomial. We have

$$(t_1 - t_2)^2 (t_2 - t_3)^2 (t_3 - t_1)^2 = \begin{vmatrix} 1 & 1 & 1 \\ t_1 & t_2 & t_3 \\ t_1^2 & t_2^2 & t_3^2 \end{vmatrix} \begin{vmatrix} 1 & t_1 & t_1^2 \\ 1 & t_2 & t_2^2 \\ 1 & t_3 & t_3^2 \end{vmatrix}$$

$$= \begin{vmatrix} 3 & t_1 + t_2 + t_3 & t_1^2 + t_2^2 + t_3^2 \\ t_1 + t_2 + t_3 & t_1^2 + t_2^2 + t_3^2 & t_1^3 + t_2^3 + t_3^3 \\ t_1^2 + t_2^2 + t_3^2 & t_1^3 + t_2^3 + t_3^3 & t_1^4 + t_2^4 + t_3^4 \end{vmatrix}.$$

Since in the present case $t_1 + t_2 + t_3 = 0$, it follows that

$$t_1^2 + t_2^2 + t_3^2 = -2(t_1 t_2 + t_2 t_3 + t_3 t_1) = -2a,$$
$$t_1^3 + t_2^3 + t_3^3 = 3 t_1 t_2 t_3 = -3b,$$
$$t_1^4 + t_2^4 + t_3^4 = 2(t_1 t_2 + t_2 t_3 + t_3 t_1)^2 = 2a^2.$$

Hence

$$D(a, b) = \begin{vmatrix} 3 & 0 & -2a \\ 0 & -2a & -3b \\ -2a & -3b & 2a^2 \end{vmatrix} = -4a^3 - 27b^2.$$

Problem 18. For which values of the parameter a do the polynomials $x^2 + 2x + a$ and $x^2 + ax + 2$ have a common real root?

Answer: $a = -3$.

Problem 19. What condition on the coefficients of the polynomials $x^2 + a_1 x + a_2$ and $x^2 + b_1 x + b_2$ is equivalent to their having at least one root in common (in the field of complex numbers)?

If ξ is a root common to both polynomials, then the vector $(\xi^2, \xi, 1)$ is a solution of the system

$$x + a_1 y + a_2 z = 0,$$

$$x + b_1 y + b_2 z = 0,$$

so that it is parallel to the vector $(a_1 b_2 - a_2 b_1, a_2 - b_2, b_1 - a_1)$, whence the condition $(a_2 - b_2)^2 = (a_1 b_2 - a_2 b_1)(b_1 - a_1)$.

The simple-minded approach to finding the analogous condition for pairs of polynomials of higher degrees meets with considerably greater difficulty. The condition is best couched in terms of the "resultant" of the two polynomials, defined as follows: The *resultant* $R_{f,g}$ of two polynomials

$$f(x) = a_0 x^n + a_1 x^{n-1} + \cdots + a_n,$$
$$g(x) = b_0 x^k + b_1 x^{k-1} + \cdots + b_k,$$

of positive degrees n and k, is the following determinant of size $n + k$, with the first k rows each having as entries the coefficients of the nth–degree polynomial f (supplemented by $k - 1$ zeros), and each of the last n rows having the coefficients of g (and $n - 1$ zeros) as entries:

$$R_{f,g} := \begin{vmatrix} a_0 & a_1 & \cdots & a_n & 0 & \cdots & 0 \\ 0 & a_0 & a_1 & \cdots & a_n & \cdots & 0 \\ \vdots & \vdots & \ddots & \ddots & \vdots & \ddots & \vdots \\ 0 & \cdots & \cdots & \cdots & \cdots & \cdots & a_n \\ b_0 & \cdots & b_k & 0 & \cdots & \cdots & 0 \\ \vdots & \ddots & \ddots & \ddots & \ddots & \ddots & \vdots \\ 0 & \cdots & \cdots & \cdots & \cdots & \cdots & b_k \end{vmatrix}.$$

Theorem 7. *The following assertions hold:*

(1) $R_{f,g} = 0$ *if and only if the polynomials* f *and* g *have a common factor of positive degree over any field containing the coefficients of* f *and* g.

(2) $R_{f,g} = a_0^k b_0^n \prod_{\substack{i=1,\dots,n \\ j=1,\dots,k}} (\alpha_i - \beta_j)$, *where* $\alpha_1, \alpha_2, \dots, \alpha_n$ *are the roots of the polynomial* f, *and* $\beta_1, \beta_2, \dots, \beta_k$ *are the roots of* g. *(Here, as before, each root is of course understood to appear as many times as its multiplicity.)*

(3) $R_{f,f'} = a_0(-1)^{n(n-1)/2} D_n$. *(Here* f' *is the derivative of* f, *and the discriminant* D_n *in the case where the leading coefficient* a_0 *of* f *is different from 1, is taken to be* $a_0^{2n-2} \Delta_n^2$.)

For (1), suppose first that f and g have a common factor h of positive degree; then the polynomials f_1 and g_1 given by $f = h f_1$, $g = -h g_1$ satisfy $\deg f_1 < n$, $\deg g_1 < k$, and $f g_1 + f_1 g = 0$.

Exercise. Prove the converse assertion, i.e., that if there exist nonzero polynomials f_1 and g_1 with $\deg f_1 < n$, $\deg g_1 < k$ satisfying $f g_1 + f_1 g = 0$, then f and g have a common factor of positive degree.

Now write $f_1 = c_0 x^{n-1} + \cdots + c_{n-1}$, $g_1 = d_0 x^{k-1} + \cdots + d_{k-1}$. On writing down the coefficients of $f g_1 + f_1 g$ and equating them to zero, one obtains a (homogeneous) linear system in the unknowns $c_0, \ldots, c_{n-1}, d_0, \ldots, d_{k-1}$ whose coefficient matrix has determinant equal to the resultant of f and g. Hence the resultant is zero if and only if there exist nonzero polynomials f_1, g_1 with the above properties.[1]

For 2), we begin with an exercise.

Exercise. Prove that

$$
\begin{pmatrix}
a_0 & a_1 & \cdots & a_n & 0 & \cdots & 0 \\
0 & a_0 & \cdots & \cdots & a_n & \cdots & 0 \\
\vdots & \vdots & \ddots & \ddots & \vdots & \ddots & \vdots \\
0 & \cdots & \cdots & \cdots & \cdots & \cdots & a_n \\
b_0 & \cdots & b_k & 0 & \cdots & \cdots & 0 \\
\vdots & \ddots & \ddots & \ddots & \ddots & \ddots & \vdots \\
0 & \cdots & \cdots & \cdots & \cdots & \cdots & b_k
\end{pmatrix}
\begin{pmatrix}
\beta_1^{n+k-1} & \cdots & \alpha_n^{n+k-1} \\
\vdots & \ddots & \vdots \\
\beta_1 & \cdots & \alpha_n \\
1 & \cdots & 1
\end{pmatrix}
$$

$$
=
\begin{pmatrix}
\beta_1^{k-1} f(\beta_1) & \cdots & \beta_k^{k-1} f(\beta_k) & 0 & \cdots & 0 \\
\vdots & \ddots & \vdots & \vdots & \ddots & \vdots \\
f(\beta_1) & \cdots & f(\beta_k) & 0 & \cdots & 0 \\
0 & \cdots & 0 & \alpha_1^{n-1} g(\alpha_1) & \cdots & \alpha_n^{n-1} g(\alpha_n) \\
\vdots & \ddots & \vdots & \vdots & \ddots & \vdots \\
0 & \cdots & 0 & g(\alpha_1) & \cdots & g(\alpha_n)
\end{pmatrix}.
$$

From this we see that

$$
R_{f,g} \prod_{i,j}(\beta_j - \alpha_i) \prod_{i>j}(\beta_j - \beta_i) \prod_{i>j}(\alpha_j - \alpha_i)
$$

$$
= \prod_j f(\beta_j) \prod_i g(\alpha_i) \prod_{i>j}(\beta_j - \beta_i) \prod_{i>j}(\alpha_j - \alpha_i).
$$

---

[1] *Translator's note.* Here is an alternative proof of part (1) of Theorem 7. It follows (as for $\mathbb{Z}$) from the existence of a "division algorithm" in the ring $F[x]$ of polynomials over any field F that $F[x]$ is a "principal ideal domain," meaning in particular that every ideal consists of all multiples hk of a particular polynomial k by arbitrary polynomials h. (See Section 5.6 below.) (As a consequence one has, analogously to $\mathbb{Z}$, that every nonzero element of $F[x]$ can be decomposed, essentially uniquely, as a product of prime polynomials of $F[x]$.)

Exercise. Deduce that two nonzero polynomials $f, g \in F[x]$ have a common factor of positive degree over some extension field of F if and only if they have a common factor of positive degree in $F[x]$.

Hence f and g have a common root in some extension field of F if and only if they have a common factor of positive degree in $F[x]$. (One needs here the fact that for any polynomial over F there is an extension field containing the full complement of roots of that polynomial, i.e., over which the polynomial factors as a product of degree–one factors.) Assertion (1) of the theorem therefore follows from (2).

In the case where the numbers $\alpha_1, \ldots, \alpha_n, \beta_1, \ldots, \beta_k$ are all distinct, this yields

$$R_{f,g} \prod_{i,j} (\beta_j - \alpha_i) = \prod_j f(\beta_j) \prod_i g(\alpha_i),$$

from which in turn, using

$$f(\beta_j) = a_0 \prod_{i=1}^{n} (\beta_j - \alpha_i), \ \ g(\alpha_i) = b_0 \prod_{j=1}^{k} (\alpha_i - \beta_j),$$

we obtain

$$R_{f,g} \prod_{i,j} (\beta_j - \alpha_i) = a_0^k b_0^n \prod_{i,j} (\beta_j - \alpha_i) \prod_{i,j} (\alpha_i - \beta_j),$$

whence the desired expression for $R_{f,g}$.

We now turn to the general case, where the numbers $\alpha_1, \ldots, \alpha_n, \beta_1, \ldots, \beta_k$ need not all be distinct. Each side of the equation in question, namely the resultant $R_{f,g}$ and the expression $a_0^k b_0^n \prod_{i,j} (\alpha_i - \beta_j)$, may be regarded as a (polynomial) function on $\mathbb{R}^{n+k}$ in $\alpha_1, \ldots, \alpha_n, \beta_1, \ldots, \beta_k$ considered as variables (taking a_0, b_0 as given, but the other coefficients of f and g as determined by these and the α_i and β_j). Since the subset of $\mathbb{R}^{n+k}$ consisting of those points $(x_1, \ldots, x_{n+k})$ with all $n + k$ coordinates distinct is everywhere dense in $\mathbb{R}^{n+k}$, and as we have already shown, the two functions coincide on this subset, they must, being certainly continuous, coincide everywhere on $\mathbb{R}^{n+k}$.

For part (3) of the theorem, observe first that by the formula just established,

$$R_{f,f'} = a_0^{n-1} \prod_{i=1}^{n} f'(\alpha_i).$$

Now, since

$$f'(x) = a_0 \sum_{j=1}^{n} \left(\prod_{i \neq j} (x - \alpha_i) \right),$$

we have $f'(\alpha_i) = a_0 \prod_{j \neq i} (\alpha_i - \alpha_j)$, whence

$$R_{f,f'} = a_0^{2n-1} \prod_{i=1}^{n} \left(\prod_{j \neq i} (\alpha_i - \alpha_j) \right)$$

$$= a_0 (-1)^{n(n-1)/2} a_0^{2n-2} \prod_{j < i} (\alpha_i - \alpha_j)^2 = a_0 (-1)^{n(n-1)/2} D_n. \ \ \square$$

Exercise. Show that the resultant of the quadratic trinomials $f(x) = a_0 x^2 + a_1 x + a_2$ and $g(x) = b_0 x^2 + b_1 x + b_2$ is

$$R = (a_0 b_2 - a_2 b_0)^2 - (a_1 b_2 - a_2 b_1)(a_0 b_1 - a_1 b_0).$$

5.6   The method of elimination and Bézout's theorem

Problem 20. Find an integer polynomial of least degree having as one of its roots:
(a) $\sqrt{2} + \sqrt{3}$; (b) $\sqrt{2} + \sqrt[3]{2}$.

Solution of (a): Setting $t := \sqrt{2} + \sqrt{3}$, one has $t^2 - 2t\sqrt{2} + 2 = 3$, whence $t^2 - 1 = 2t\sqrt{2}$, whence in turn $t^4 - 10t^2 + 1 = 0$.

Answer to (b): $t^6 - 6t^4 - 4t^3 + 12t^2 - 24t - 4$.

A general approach to solving this sort of problem will be expounded below. In the meantime here is another problem.

Problem 21. Prove that if two quadratic plane curves have only finitely many points in common, then there are at most four such points.

We give two solutions. The first uses no theoretical considerations whatsoever. Consider the system

$$\begin{cases} P(x, y) := a_{11}x^2 + 2a_{12}xy + a_{22}y^2 + 2a_{13}x + 2a_{23}y + a_{33} = 0, \\ Q(x, y) := b_{11}x^2 + 2b_{12}xy + b_{22}y^2 + 2b_{13}x + 2b_{23}y + b_{33} = 0, \end{cases}$$

made up of the equations of the two curves. Since the difference $b_{11}P(x, y) - a_{11}Q(x, y)$ has no term in x^2, we can solve the equation $b_{11}P(x, y) - a_{11}Q(x, y) = 0$ for x in terms of y, obtaining, say, $x = h_2(y)/h_1(y)$, where $\deg h_2 \leq 2$ and $\deg h_1 \leq 1$. Substituting this expression for x in either equation and multiplying throughout by h_1^2, we obtain an equation in y of degree at most 4.

Here is the second solution: First express the polynomials P and Q in terms of powers of x:

$$P(x, y) = p_0(y)x^2 + p_1(y)x + p_2(y), \quad Q(x, y) = q_0(y)x^2 + q_1(y)x + q_2(y).$$

If $P(x_0, y_0) = Q(x_0, y_0) = 0$, then the trinomials $P(x, y_0)$ and $Q(x, y_0)$ have a root in common, so that their resultant vanishes. Hence y_0 is a root of the polynomial

$$(p_0(y)q_2(y) - p_2(y)q_0(y))^2$$
$$- (p_1(y)q_2(y) - p_2(y)q_1(y))(p_0(y)q_1(y) - p_1(y)q_0(y)),$$

which has degree at most four. Hence there are at most four possible y-coordinates of points of intersection, and similarly for the x-coordinates. It remains only to observe that by means of a rotation of axes one can arrange that no vertical line $x = y_0$ has on it more than one of the points of intersection of the curves.

Theorem 8 (Bézout).   *Let $f(x, y)$ and $g(x, y)$ be polynomials of degrees n and k respectively. If they have no common divisor of positive degree, then the system*

$$\begin{cases} f(x, y) = 0, \\ g(x, y) = 0 \end{cases}$$

has at most nk solutions.

Corollary. *A quartic curve cannot consist of more than four ovals.*

For suppose on the contrary that there is a curve $p_4(x, y) = 0$ of degree at most four consisting of four ovals and at least one more point M, say. It follows from the argument given at the end of Section 5.2 that none of these ovals can lie inside any other. Choose points A, B, C, D, one inside each oval, and let $q_2(x, y) = 0$ be a curve of degree at most two passing through these points together with M. It is then geometrically clear that this curve meets the given curve in at least nine points, contradicting Bézout's theorem. $\square$

Exercise. Prove that provided the number $\varepsilon > 0$ is sufficiently small, the curve defined by the equation $2x^4 + 2y^4 + 5x^2y^2 - 3x^2 - 3y^2 + 1 + \varepsilon = 0$ consists of four separate pieces.

(Use $2x^4 + 2y^4 + 5x^2y^2 - 3x^2 - 3y^2 + 1 = (x^2 + 2y^2 - 1)(2x^2 + y^2 - 1)$. We are not yet equipped to show that the four pieces are actually ovals; see Section 9.7.)

We leave the proof of Bézout's theorem to the reader as an exercise: the idea of the proof is as illustrated in the second solution of Problem 21. Instead of simply going through that argument again, albeit in its general form, we give another solution of Problem 20 (a).

Write $x := \sqrt{2}$, $y := \sqrt{3}$ and $z := x + y = \sqrt{2} + \sqrt{3}$. Then from $x^2 = 2$, $y^2 = 3$, $x + y = z$, we obtain $(z - x)^2 = x^2 - 2xz + z^2 = 3$. Hence the polynomials (in x) $x^2 - 2$ and $x^2 - 2xz + (z^2 - 3)$ have a root in common. The resultant of these two polynomials is

$$
\begin{vmatrix} 1 & 0 & -2 & 0 \\ 0 & 1 & 0 & -2 \\ 1 & -2z & z^2 - 1 & 0 \\ 0 & 1 & -2z & z^2 - 3 \end{vmatrix} = \begin{vmatrix} 1 & 0 & -2 \\ -2z & z^2 - 1 & 0 \\ 1 & -2z & z^2 - 3 \end{vmatrix}
$$

$$
= (z^2 - 1)(z^2 - 3) - 8z^2 + 2(z^2 - 1) = z^4 - 10z^2 + 1.
$$

Since this must vanish, we have found a polynomial over $\mathbb{Z}$ having z as a root.

The general result may be formulated as follows:

Theorem 9. *Let $f(x)$, $g(y)$, and $h(x, y)$ be arbitrary rational polynomials. If α and β are roots of f and g respectively, then the number $\gamma := h(\alpha, \beta)$ is also a root of some rational polynomial p.[2]*

Such a polynomial p can be constructed by means of "elimination" of x, y from the system

$$
f(x) = 0,
$$
$$
g(y) = 0,
$$

---

[2] *Translator's note.* More generally: For any field extension $k \subset K$, those elements of K that are roots of polynomials over k, i.e., that are "algebraic" over the ground field k, as they say, form a subfield of K.

$$z = h(x, y),$$

as just illustrated. On the other hand, if one is satisfied with establishing the theorem as stated, i.e., with merely showing the existence of some such polynomial p, then one may proceed otherwise.

We take as our universe the field $\mathbb{C}$ of complex numbers (and for definiteness take $\mathbb{Q}$ as the ground field).

Lemma 3. *If α is a root of some rational polynomial (i.e., an "algebraic number"), then the set*

$$\mathbb{Q}[\alpha] := \{g(\alpha) \mid g \in \mathbb{Q}[t]\},$$

is a subfield of $\mathbb{C}$.

That $\mathbb{Q}[\alpha]$ is closed under addition, subtraction, and multiplication is clear. Thus we need only show that if $0 \neq \beta = g(\alpha) \in \mathbb{Q}[\alpha]$, then $1/\beta \in \mathbb{Q}[\alpha]$. To this end, let $d \in \mathbb{Q}[t]$ be a polynomial of smallest degree having α as a root. Dividing d into g (using the "division algorithm"), we obtain $g(t) = q(t)d(t) + r(t)$, where the remainder satisfies $r(t) \equiv 0$ or $\deg r < \deg d$. Substitution of α for t yields $\beta = g(\alpha) = q(\alpha)d(\alpha) + r(\alpha) = r(\alpha) \neq 0$. Hence r is a nonzero polynomial of degree $< \deg d$.

Exercise. Prove that the polynomial d (the "minimal polynomial" of α) is irreducible over $\mathbb{Q}$, i.e., is not expressible as a product of two rational polynomials of positive degree.

(In this context "irreducible" $\equiv$ "prime.")

Hence since r has smaller degree than d, these polynomials must be relatively prime. It follows (from the fact that $\mathbb{Q}[t]$ is a principal–ideal domain—see the proof of part (1) of Theorem 7 above) that there exist polynomials $A, B \in \mathbb{Q}[t]$ such that $Ad + Br = 1$, whence $A(\alpha)d(\alpha) + B(\alpha)r(\alpha) = 1$, i.e., $B(\alpha)r(\alpha) = 1$. Thus the inverse $B(\alpha)$ of β is in $\mathbb{Q}[\alpha]$. $\square$

Exercise. Prove that as a vector space over $\mathbb{Q}$, the field $\mathbb{Q}[\alpha]$ has dimension $\dim_{\mathbb{Q}} \mathbb{Q}[\alpha] = \deg d$.

In the proof of Lemma 3 no use was made of the fact that the ground field was the particular field $\mathbb{Q}$. Essentially the same argument shows that for any field extension $k \subset K$ whatever, if $\alpha \in K$ is a root of some polynomial over the ground field k, then $k[\alpha] := \{g(\alpha) \mid g \in k[t]\}$ is a subfield of K (containing k). Hence, in particular, if α, β are roots of the given rational polynomials f, g respectively, then

$$\mathbb{Q}[\alpha, \beta] := \{s(\alpha, \beta) \mid s \in \mathbb{Q}[t, u]\} = \{q(\beta) \mid q \in \mathbb{Q}[\alpha][t]\} = \mathbb{Q}[\alpha][\beta]$$

is again a field.

Lemma 4. *The field $\mathbb{Q}[\alpha, \beta]$ has finite dimension as a vector space over $\mathbb{Q}$.*

Observe that $\mathbb{Q}[\alpha, \beta]$ has finite dimension over $\mathbb{Q}[\alpha]$, which in turn has finite dimension over $\mathbb{Q}$. Hence . . . .

Exercise. Complete the proof of the lemma.

We are now ready to complete the proof of Theorem 9. We have $\gamma := h(\alpha, \beta) \in \mathbb{Q}[\alpha, \beta]$, a field of finite dimension n, say, over $\mathbb{Q}$. Hence the $n + 1$ elements $1, \gamma, \ldots, \gamma^n$ are linearly dependent over $\mathbb{Q}$, i.e., there exist rational numbers $a_i \in \mathbb{Q}$, not all zero, such that $\sum_{i=0}^{n} a_i \gamma^i = 0$, so that we may take $p(t) = \sum_{i=0}^{n} a_i t^i$. $\square$

Exercise. Show that Theorem 9 remains valid for any *rational* function $h(x, y)$ over $\mathbb{Q}$.

5.7 The factor theorem and finite fields

In this final section we shall prove the following interesting result, which moreover we shall need in Chapter 7.

Theorem 10. *The multiplicative group $F^* := F \setminus \{0\}$ of a finite field F is cyclic.*

Let R be an integral domain, i.e., a commutative ring with a multiplicative identity $1 \neq 0$, in which there are no zero–divisors. The *characteristic* p of R is the natural number defined by $p := \min\{k \in \mathbb{N} \mid k \times 1 = 0\}$ if this set is nonempty; otherwise it defined to be zero. (Observe that here $k \times 1$ is not a product in R, but merely shorthand for $1 + 1 + \cdots + 1$ (k times); thus the equation $p \times 1 = 0$ does not contradict the assumption that there are no zero–divisors in the ring R.)

Exercise. Prove that the characteristic of an integral domain is either zero or a prime. Deduce that the characteristic of a finite integral domain (which must in fact be a field—why?) is a prime.

Lemma 5. *If R is a finite field of characteristic p, then $|R| = p^n$ for some $n \in \mathbb{N}$.*[3]

It is not difficult to see that by neglecting part of its multiplicative structure, an integral domain of finite (prime) characteristic may be considered as a vector space over the field of p elements $\mathbb{Z}/p\mathbb{Z}$. If the integral domain is finite, then the dimension of this vector space must be finite. $\square$

We shall now look a little more closely at the division algorithm for polynomials.

Problem 22. Let $P(x)$ be a polynomial such that the remainders after division of $P(x)$ by $x - 2$ and $x + 1$ are respectively 3 and 1. Find the remainder after division by $x^2 - x - 2$.

By assumption $P(x) = (x - 2)q_1(x) + 3$ and also $P(x) = (x + 1)q_2(x) + 1$. Setting $x = 2$ and $x = -1$ in turn yields $P(2) = 3$ and $P(-1) = 1$. Division by $x^2 - x - 2$ gives

$$P(x) = (x^2 - x - 2)q(x) + ax + b = (x - 2)(x + 1)q(x) + ax + b,$$

---

[3] *Translator's note.* It can be shown that in fact for each natural number of the form $p^n > 1$ (p prime), there is up to isomorphism exactly one field of that order.

whence $2a + b = P(2) = 3$ and $-a + b = P(-1) = 1$. Thus the answer is obtained by solving the system

$$2a + b = 3,$$
$$-a + b = 1.$$

The procedure ("division algorithm") for dividing one polynomial into another by "long division" is essentially the same whatever the coefficient field, so that as noted earlier, many of the basic results concerning polynomials with numerical coefficients hold more generally for polynomials over any field. In particular, the factor theorem is valid for polynomials in $F[x]$, F any field, i.e., $a \in F$ is a root of $f \in F[x]$ if and only if $x - a$ divides f. Consequently, a polynomial of degree n in $F[x]$ has at most n roots in F (in fact, in any extension field $K \supset F$). (Theorem 4 above is just this assertion, although under the implicit assumption that F is a subfield of $\mathbb{C}$; however, the proof of that theorem goes through without change whatever the field of provenance of the coefficients of the polynomial in question.)

For our present purposes, F will be a finite field of characteristic p. Thus p is prime, and by Lemma 5, $|F| = p^n$ for some natural number n. Set $q := p^n - 1$.

Lemma 6. *For any natural number r dividing q, the polynomial $x^r - 1$ has exactly r distinct roots; moreover, all are in F.*

Since the multiplicative group F^* has order q, its elements all satisfy the equation $x^q = 1$. Hence the polynomial $x^q - 1$ has its full complement of q (distinct) roots in F^*. Since r divides q, we have $x^q - 1 = (x^r - 1)h(x)$, where $\deg h = q - r$. Since at most $q - r$ of the q distinct roots of $x^q - 1$ can come from $h(x)$, it follows that $x^r - 1$ must have exactly r distinct roots, all in F^*. $\square$

Theorem 10 is now immediate from the following group-theoretical result.

Lemma 7. *Let G be an abelian group of order q. If for each natural number r dividing q, G has exactly r elements g satisfying $g^r = 1$, then G is cyclic.*

First express q as a product of powers of distinct primes: $q = p_1^{m_1} \cdots p_k^{m_k}$. It follows from the assumption of the theorem that

$$\forall i = 1, \ldots, k, \; \exists g_i \in G : \; g_i^{p_i^{m_i}} = 1, \; g_i^{p_i^{m_i-1}} \neq 1.$$

Clearly then, each g_i must have order $p_i^{m_i}$.

Exercise. Show that if g, h are commuting elements of a group, of relatively prime orders s, t respectively, then gh has order st.

(Prove first that gh has order dividing st, and hence of the form $s_1 t_1$ where s_1 divides s and t_1 divides t. Next, suppose for instance that $s_1 < s$. Then

$$1 = (gh)^{s_1 t} = g^{s_1 t} h^{s_1 t} = g^{s_1 t} = \left(g^t\right)^{s_1},$$

so that g^t has order dividing s_1. However, from the fact that $us + vt = 1$ for appropriate integers u, v, it follows readily that g^t has the same order s as g.)

Returning to the proof of the lemma, we have that $g_1 g_2$ has order $p_1^{m_1} p_2^{m_2}$, and then that $g_1 g_2 g_3$ has order $p_1^{m_1} p_2^{m_2} p_3^{m_3}$, and so on. $\square$

The approach to learning how to solve equations and especially inequalities, involving representation of the solution sets in the plane, has many advantages from the teacher's point of view. For then a student can see his or her mistakes all the more clearly: since so many radically different pictures are possible, a difference due to error is visually striking. In contrast with the more common equations and inequalities, where often a greater degree of challenge is bought at the expense of introducing cumbersome expressions into their formulation, even very simple–looking inequalities "in x and y" (easy to invent in unlimited numbers) often have unexpected and appealing graphs. Essentially no new methods are needed beyond those used in solving equations and inequalities on the line, and furthermore—and this is very important—in drawing the solution sets, the students are compelled unwittingly to do a certain amount of reasoning, in addition to merely manipulating formulae. Here there are actually very few logical ideas involved; the basic ones are given in Theorems 1 and 3. Finally, it might be noted that here there is ample scope for the teacher to make up series of problems.

Theorem 2 calls for some commentary. It is a trivial consideration, which nonetheless is sometimes forgotten, that if the graph of a function f is regarded purely geometrically as a subset of the plane (defined by the equation $y = f(x)$, or $y - f(x) = 0$), then there is no essential difference between, for example, the translations of the graph through a distance a in the direction of the x-axis or the y–axis. In both cases the corresponding equation is obtained from the original by subtracting a from the appropriate variable, obtaining in the first case $y - f(x - a) = 0$, i.e., $y = f(x - a)$, and in the second $y - a - f(x) = 0$, or $y = f(x) + a$! It is in general essential to preserve this sort of "functional" approach to the solution of equations and inequalities, rather than the purely manipulative algebraic one where one hopes to be able to apply a standard formula.

By this no undue deprecation of elementary algebra is intended. No method develops a "feel for formulae" or inculcates the skill of unraveling complicated formulae as thoroughly as practice in factoring algebraic expressions. And of course how many interesting problems there are involving such simple algebraic formulae as, for instance, Viète's formula (expressing the coefficients of a polynomial in terms of the roots)! Note in conclusion the modest slice of "modern algebra" served up in the final two sections of the chapter.

CHAPTER 6

Graphs

6.1 Graphical reformulations

We begin with three problems.

Problem 1. Can one by means of a sequence of standard knight–moves get from the configuration of knights shown in Diagram a to that of Diagram b? (It is intuitively clear that this is impossible, but how can one clinch this with a proof?)

One may proceed as follows: First number the nine squares from 1 to 9 as shown in Diagram c, and then label nine points in the plane also with the numbers 1 to 9; each of these is to represent the square with the same number.

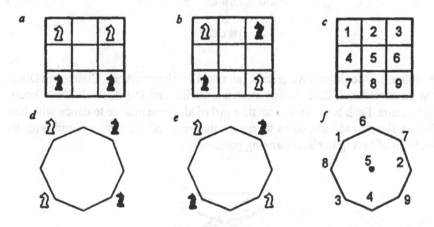

If two squares are such that a knight can get from one to the other in a single move, join the corresponding two points by means of an arc (avoiding the other points);

this results in a figure essentially like that of Diagram f. The given initial and final arrangements of knights are then represented by Diagrams d and e respectively. Since the initial arrangement has the two white knights joined by a path without obstacles, while in the final one the black knights intervene, it is indeed impossible to get from one arrangement to the other in the prescribed manner.

Problem 2. A traveler with a wolf, a goat, and a cabbage arrives at a river's edge. On the bank there is a raft, which however will bear at most the traveler and one other without sinking. Naturally, the traveler cannot leave unsupervised the wolf with the goat or the goat with the cabbage. How can he transport himself and his possessions to the opposite bank without risking any loss?

Although it is not difficult to do this problem in one's head, we shall show how to solve it using a certain graph. With each "state" in the process of transporting wolf, goat, and cabbage to the opposite bank, we associate a point, labeled appropriately; thus, for example, the point representing the initial state, where none of the traveler's three possessions has yet been ferried over, will be labeled $WGC|$ (see the diagram). These points will be the vertices of the graph in question. As shown in the diagram, two of these points are then joined by an edge if the traveler can change either one of the corresponding states into the other by ferrying one of his possessions across the river, at no risk to goat or cabbage. The solutions are then given by the paths leading from the vertex $WGC|$ to the vertex $|WGC$. (One such is shown in boldface in the diagram.)

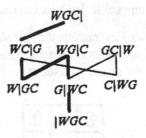

Problem 3. Alec knows Anna and Claire, and Bob knows Anna, Claire, and Diana. On the other hand, Chris knows Beatrice, Claire, and Diana, while Dave knows only Claire. Each boy wants to invite a girl of his acquaintance to dance with him. Whom should each ask so as to ensure that none of the girls is faced with the problem of having to choose among partners?

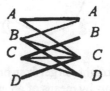

The solution is shown in the diagram.

In solving each of these three problems it turned out to be useful to first find a graph "picturing" the conditions assumed in the problem, and then to translate the question into one concerning this graph.

Here is the precise definition of a "graph." The formal concept best suited to our purposes, and perhaps most commonly encountered, is that of a *simple unoriented graph*: this is defined to be a pair V, E where V is an arbitrary nonempty set and E is any set of unordered pairs $\{v, v'\}$ of elements of V with $v \neq v'$. The elements of the set V are called the *vertices* of the graph, and those of E its *edges*. We say of each edge (v, v') that it *joins* v and v' (or *goes* from v to v', or, equally, from v' to v) in the given graph.

In what follows we shall not in fact use such formal terminology, but shall consider a graph G to be made up simply of a set V of vertices, and a set E of edges each joining a unique pair of distinct vertices. The term *simple* means that there are no "loops," i.e., edges joining a vertex to itself, and no "multiple edges," i.e., there is at most one edge joining any pair of vertices.

The *valency* (or *degree*) $\rho(v)$ of a vertex v is defined to be the number of edges incident with v. A (finite) *path* in a graph is given by a sequence $(v_0, \ell_0, \ldots, \ell_k, v_{k+1})$ of vertices and edges, where the edge ℓ_i joins the vertices v_i and v_{i+1}, $i = 1, \ldots, k$. For the most part we shall be considering only finite graphs, i.e., graphs where the sets V and E are both finite.

Here is another easy problem.

Problem 4. In the country known as "Alphabet," there are ten towns: $A, B, \ldots, J$ linked by roads as follows: There is a road from A to B, from F to D, C to E, I to B, H to J, G to E, F to A, J to C, I to D, and H to G. Is it possible to travel between the towns A and J using these roads?

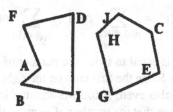

On drawing the graph with vertices representing the towns, and edges the roads (see the diagram), one sees clearly that it is not possible to go from A to J via those roads.

We conclude this introductory section with a logical problem not easily solved in one's head.

Problem 5. The following facts are known:

A. No fish without at least three rows of teeth feels sure that it is adequately armed.

B. All sharks feel confident of being well armed.

C. Any fish not able to dance the quadrille deserves sympathy.
D. All fish except sharks are always kind to children.
E. Big fish are not able to dance the quadrille.
F. Any fish with at least three rows of teeth does not deserve any sympathy.
Does it then follow that big fish are always kind to children?

(Draw a directed graph with edges corresponding to the assertions A to F, on the following model of statement F:
A fish has at least three rows of teeth $\longrightarrow$ it deserves no sympathy.)

6.2   Graphs and parity

What do the following two problems have in common?

Problem 6. Can 1001 telephones be interconnected in such a way that each is connected to exactly eleven others?

Problem 7. After his trip to Disneyland, John related how in the enchanted lake there are seven islands, each of which has one, three, or five bridges leading from it. Does it follow, he asked, that at least one of these bridges exits onto the shore of the lake?

The solutions of these problems are immediate, once one has the following general result.

Theorem 1. *In any (finite) graph the number of vertices of odd valency is even.*

To see why this must be so, note first that

$$\sum_{v \in V} \rho(v) = \sum_{\rho(v) \text{ even}} \rho(v) + \sum_{\rho(v) \text{ odd}} \rho(v).$$

Now, the sum $\sum_{v \in V} \rho(v)$ is equal to twice the number of edges of the graph, and is therefore certainly even. Since the first sum on the right–hand side of the above equation is—obviously—also even, it follows that the last sum in the equation must also be even, and hence that the number of terms in that sum is even, i.e., the number of vertices with odd valency is even. $\square$

This theorem is often restated in the following appealing form: The number of people who up to the present moment have shaken hands an odd number of times is even.

Problem 8. During the twenty-seventh dynasty, when flying carpets were the usual means of travel, there were twenty-one carpet routes leading out of the imperial capital. Every other town of the empire had exactly ten such flight–paths emanating from it, excepting the settlement Faraway, which had only one. Show that Faraway must have been accessible by carpet from the capital, albeit with transfers.

If, on the contrary, Faraway were not thus accessible, then in the graph with vertices the capital together with those towns accessible from it (and the obvious edges), there would be just one vertex of odd valency.

In the solutions of Problems 4, 5 and 8 above, the notion of "connectedness" (and "connected component") of a graph was used implicitly. However, in these two problems appeal was made to two different, though equivalent, defining conditions for this concept. Although their equivalence is pictorially obvious, it still requires proof (especially in the situation of an infinite graph).

Theorem 2. *The following two statements concerning a graph are equivalent:*
(1) *Any two vertices can be joined by a path.*
(2) *For any partition of the set V of vertices into two nonempty subsets V_1 and V_2, there is an edge joining some vertex of V_1 to some vertex of V_2.*

(The implication (1) $\Longrightarrow$ (2) is easy; for the reverse implication use *reductio ad absurdum*.) $\square$

6.3 Trees

Problem 9. In a certain lagoon there are 17 islands. (a) What is the least number of bridges needed for every island to be accessible from every other? (b) Working without a plan, the local inhabitants constructed 20 bridges connecting the islands. Prove that four of these bridges can be removed while still allowing the islanders to get from any island to any other.

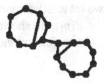

The answer to (a) would seem to be obvious: one needs at least 16 bridges. Yet how exactly does one prove that a smaller number will not suffice?

In connection with (b), note that there may be bridges that cannot be removed without affecting the connectivity (see the figure).

In solving this problem graphs of a special sort arise, namely "trees." We proceed to the definition. We first define a *chain* in a graph to be a path in which all of the edges are distinct, a *simple* chain to be a chain where in addition all of the vertices except possibly the first and last are distinct, and finally a *cycle* to be a simple chain where the first and last vertices do in fact coincide. A *tree* is then defined to be a connected graph without cycles.

The following theorem gives several equivalent conditions for a graph to be a tree, particularly in the finite case. An edge of a graph is called *critical* if its removal disconnects the graph.

Theorem 3. *Let T be a graph with n vertices. The following conditions are equivalent:*

(1) *The graph T is connected and has no cycles.*

(2) *The graph T has exactly $n - 1$ edges and no cycles.*

(3) *The graph T is connected and has exactly $n - 1$ edges.*

(4) *The graph T is connected, and every edge is critical.*

(5) *For each pair of vertices of T there is exactly one chain joining them.*

(6) *The graph has no cycles, but insertion of a new edge between any two vertices not so joined previously results in a graph with a cycle.*

We shall prove only the implication (1) $\Longrightarrow$ (2), i.e., we shall show that for trees one has the formula $|V| - |E| = 1$ relating the number of vertices and the number of edges. As we shall see later on, this represents a particular case of "Euler's formula" for planar graphs.

Lemma 1. *In a finite graph without cycles, there is a vertex of valency at most one.*

If the graph has a vertex incident with no edges, i.e., of valency zero, there is nothing to prove, so we may assume that there are no such vertices. The lemma is then easy to see intuitively; for if one imagines oneself moving along the graph without retracing one's steps, then since one cannot pass through any vertex twice, eventually one must come to a dead end, i.e., to a vertex of valency one.

Here is the formal argument. Fix on a vertex $v_0 \in V$, and with each vertex v connected to v_0 associate the length (i.e., number of edges) $d(v)$ of a shortest path from v_0 to v, the *distance*, say, between v_0 and v in the graph. Let v_1 be a vertex at greatest distance from v_0. We shall show that v_1 must have valency one. Suppose otherwise; there will then exist two (at least) distinct vertices v' and v'' adjacent to v_1 (as in the figure). Since $d(v') \leq d(v_1)$, no shortest path from v_0 to v' can pass through v_1, and similarly for v''. It now follows easily that there must be a cycle in the graph. $\square$

To prove that (1) $\Longrightarrow$ (2), i.e., that the formula $|V| - |E| = 1$ holds for any tree T, we use induction on the number $n := |V|$. The case $n = 1$ is obvious. Suppose $n > 1$ and inductively that the formula holds for trees with fewer than n vertices. By the lemma there is a vertex v_1, say, incident with just one edge ℓ_1, say, of T. Consider the subgraph T' of T with $V' := V \setminus \{v_1\}$, $E' := E \setminus \{\ell_1\}$; clearly, T' is connected and without cycles, i.e., is a tree. Hence by the inductive assumption,

$$1 = |V'| - |E'| = (|V| - 1) - (|E| - 1) = |V| - |E|.$$

Exercise. Prove the remaining implications in the theorem. □

Suppose that we have a one-to-one correspondence between the vertices of a given graph and a set of points of the plane, and that furthermore with each edge there is associated a non-self-intersecting polygonal arc in the plane joining the points corresponding to the endpoints of that edge. If these polygonal arcs do not have interior points in common, i.e., do not cross, we say that the points together with the polygonal arcs constitute a *planar realization* of the given graph, or, less formally, that that we have an *embedding* of the graph in the plane.

Exercise. Show that every tree can be embedded in the plane.

Theorem 4. *No planar realization L of a finite tree T separates the plane, i.e., any two points of the complement* $\mathbb{R}^2 \setminus L$ *can be joined by a polygonal arc avoiding L (or as they say, the set* $\mathbb{R}^2 \setminus L$ *is "pathwise connected").*

This statement may seem obvious. However, as the above diagram shows, it may be quite, or even exceedingly, difficult to find one's way out of the "labyrinth," or "maze," determined by a tree. (Incidentally, this is in fact essentially what a maze is, is it not?)

Again we use induction, this time on the number of edges of the tree. Note first that by introducing additional vertices into the tree we can arrange that in its planar realization the edges are just straight-line segments; thus we may without loss of generality take our universe of discourse to consist of finite planar trees with edges straight-line segments. The first step of the induction being obvious, consider any such tree L with $n > 0$ edges, and assume inductively that the assertion of the theorem is true for such trees with fewer than n edges. By Lemma 1 there is a vertex (point) v incident with exactly one edge (line segment) ℓ; write L' for the planar tree obtained from L by removing the vertex v and the edge ℓ. Let P, Q be any points in $\mathbb{R}^2 \setminus L'$. By the inductive assumption there is a polygonal arc $\Gamma_1 \subset \mathbb{R}^2 \setminus L'$ joining P and Q. Hence $\Gamma_1 \cap L = \Gamma_1 \cap \ell$. If any point in $\Gamma_1 \cap \ell$ is a vertex, i.e., corner point, of the polygonal arc Γ_1, we can deform Γ_1 by an arbitrarily small amount in a neighborhood of that point so as either to eliminate that point of intersection (if the polygonal arc does not actually cross ℓ there) or to ensure that while remaining a point of intersection, the point ceases to be a vertex of Γ_1; hence we may assume without loss of generality that there are no corner points of Γ_1 on ℓ.

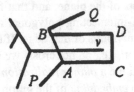

This assumed, let M denote the point of the intersection $\Gamma_1 \cap \ell$ closest to the endpoint v of the line–segment ℓ, and s the segment of the polygonal arc Γ_1 crossing ℓ at M. By replacing a sufficiently small segment AB of s by a polygonal arc $ACDB$ sufficiently close to the segment of ℓ from M to v (see the diagram), we obtain a new polygonal arc Γ_2 with one less point of intersection with L. Continuing in this way, we eventually obtain a polygonal arc from P to Q avoiding L. (Of course, strictly speaking, a further induction is required, here on the number of points in $\Gamma_1 \cap \ell$.) $\square$

6.4   Euler's formula and the Euler characteristic

Henceforth we shall for simplicity talk of planar graphs, graphs on the sphere, etc.

Theorem 5 (Euler).   *Let G be a finite, connected, planar graph ($G \subset \mathbb{R}^2$), and F the set of connected components of the complement $\mathbb{R}^2 \setminus G$. The following formula holds:*

$$|V| - |E| + |F| = 2.$$

We shall use induction on the number of edges of G. Thus, proceeding to the inductive step, suppose that G has at least one edge and, inductively, that Euler's formula holds for planar graphs with fewer edges. If G is a tree, then by Theorem 4 we have $|F| = 1$, and writing the formula for finite trees proved in the preceding section in the form $|V| - |E| + 1 = 2$, we see that Euler's formula holds for trees. Hence we may suppose that G is not a tree, whence by Theorem 3 it has a noncritical edge ℓ, say. Thus the graph G' obtained by removing this edge remains connected, and so by the inductive assumption the formula $|V'| - |E'| + |F'| = 2$ holds for it. It now only remains to observe first that $|E'| = |E| - 1$, and second that since the edge ℓ separates one of the components of $\mathbb{R}^2 \setminus G'$ into two, we have $|F'| = |F| - 1$. $\square$

Exercise. Generalize Euler's formula to the situation of an arbitrary finite planar graph.

Problem 10. Each of three houses is to be connected up by utility lines to the water, gas, and electricity mains. Can this be done without any of the nine utility lines crossing?

Giving a completely rigorous solution to this problem would take some time, since we should first have to define a (continuous) "curve," or arc thereof, and then.... We might instead assume that the pipelines and cables take the form of polygonal arcs. However, it is more natural to rest content with an intuitive understanding of what a continuous curve is, and to take on trust that Euler's formula continues to hold when the edges of a (geometrically realized) graph are allowed to be arcs, i.e., segments, of such curves.

Thus the question is the following: Is there a planar realization of the graph $G_{3,3}$ shown in the diagram?

Here we have $|V| = 6$ and $|E| = 9$. If there were a planar realization of this graph, then each of the regions into which this realization subdivided the plane would have to be at least four-sided, i.e., have at least four edges in its boundary. Also, by Euler's formula we should have that the number of such regions is 5, i.e., $|F| = 5$. Since in the present case each edge would have to be part of the boundary of two distinct regions, it follows that there would have to be at least $4 \cdot 5/2 = 10$ edges, whereas there are only 9.

Exercise. Prove that the complete graph on 5 vertices (i.e., with every vertex adjacent to every other) cannot be embedded in the plane.

By way of an giving another interesting application of Euler's formula, we shall use it to prove the following famous result.

Theorem 6. *There exist at most five regular polyhedra.*

Lemma 2. *Euler's formula holds also for graphs on the sphere.*

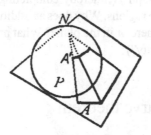

Fix on any point $N \in S^2 \setminus G$, and denote by P the diametrically opposite point of the sphere. Consider the plane tangent to the sphere at P. Then with each point A of the plane we associate the point A' in which the line segment NA meets the sphere (see the diagram). (This is the "stereographic projection" in reverse.) This defines a one-to-one (and bicontinuous) correspondence between the plane and the sphere with the point N removed. Hence an arbitrary graph G on the sphere is mapped under this bijection to a graph G' in the plane (and conversely), and clearly $|V| = |V'|, |E| = |E'|, |F| = |F'|$. $\square$

We now prove Theorem 6. Suppose that we have a regular polyhedron in which each face is an n-gon, and each vertex is incident with q edges. If we project the polyhedron onto its circumscribed sphere from any point inside the polyhedron, we obtain a connected graph on the sphere for which $n|F| = 2|E| = q|V|$. Hence Euler's formula gives

$$2\frac{|E|}{q} - |E| + 2\frac{|E|}{n} = 2, \text{ or } \frac{1}{q} + \frac{1}{n} = \frac{1}{|E|} + \frac{1}{2}.$$

n	3	3	4	3	5
q	3	4	3	5	3

Since $n, q \geq 3$, and $1/n + 1/q > 1/2$, we have just five possibilities, as shown in the table. $\square$

(To show that these five possibilities are actually realized, one has to exhibit regular polyhedra with such n and q. The first three pairs of values correspond respectively to the familiar tetrahedron, octahedron, and cube, the remaining two to the icosahedron and dodecahedron.)

Let us for a moment consider Euler's formula from another angle. Since that formula tells us that the quantity $|V| - |E| + |F|$ is the same for every graph on the sphere, i.e., that it is invariant, it follows that the fact that this invariant is in fact equal to 2 must be a property of the sphere! (In mathematical parlance, the "Euler characteristic" of the sphere is 2.) One might naturally expect that there should be analogous invariants for other surfaces. This expectation turns out to be justified, but not in quite as simple a fashion as one might imagine at first. For instance, it is easy to see that there are two essentially different embeddings of the graph $\triangle$ onto the torus, one of which has $|F| = 2$, and the other $|F| = 1$! (One can adjoin an edge to an embedded graph $\wedge$, thereby completing the triangle, but without separating the torus into two regions. Why does an additional region always appear when this is done on the sphere or in the plane? What property of the latter spaces have we been using implicitly?)

6.5   The Jordan curve theorem

In the proof of Theorem 5 we used implicitly the following intuitively obvious fact, known as the "Jordan curve theorem":

Theorem 7 (Jordan).   *A simple, closed, continuous curve in the plane, separates the plane into two regions (connected components).*

Once again we shall avoid the technicalities involved in defining a "continuous curve," and instead prove the theorem for a simple (i.e. non-self-intersecting) polygon; in any case, if the curve is assumed to be smooth, the theorem follows relatively easily from this specialized version. (However, the proof of the theorem in full generality is considerably more difficult.)

Thus we have to prove that for any simple plane closed polygon L, one has $\mathbb{R}^2 \setminus L = U \cup V$, where U and V are the connected components of this set, i.e., any two points that are either both in U or both in V can be connected by a polygonal arc avoiding L, while if one lies in U and the other in V, this is not possible. It is intuitive that a point of the complement of L may lie "inside" or "outside" the polygon L, and that one can distinguish the two cases by means of the parity of the number of times a ray emanating from the point crosses L (see the diagram below), i.e., by the "degree modulo 2 of the point relative to the polygon L." This is a simplified version of the general mathematical concept of the "degree of a point relative to a plane curve"; the proof that we are about to give, which incidentally is due to Hilbert [14], is interesting in particular for its use of this concept, albeit in this simple form.

Let $M \notin L$, and let t be a ray emanating from M. To begin with suppose that this ray does not pass through any vertex of L; in this case one defines the *index* of the ray t relative to the polygon L by

$$\mathcal{I}(M, L; t) := \begin{cases} 0 \text{ if the number } |t \cap L| \text{ is even,} \\ 1 \text{ if the number } |t \cap L| \text{ is odd.} \end{cases}$$

(Here, as always, $|X|$ denotes the number of elements in the set X; thus $|t \cap L|$ is just the number of points of intersection of the ray with the polygon.)

If, on the other hand, the ray t passes through a vertex A of L, then obviously the disposition relative to t of the two edges of the polygon incident with A must be as shown in one of the four figures a–d below. In the first and third cases, where the polygon crosses the ray, for the purpose of calculating the index we consider there to be just one point of intersection, and in the second and fourth, none.

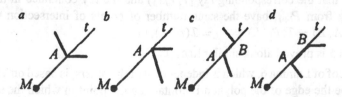

Having completed the definition of index, we establish the basic properties of it that we shall be needing.

Lemma 3. *The number $\mathcal{I}(M, L; t)$ is independent of the ray t.*

In view of this fact we may without ambiguity write $\mathcal{I}(M, L)$ for the index of M relative to L.

Lemma 4. *If points A and B can be connected by a polygonal arc avoiding the polygon L, then $\mathcal{I}(A, L) = \mathcal{I}(B, L)$.*

In the next two lemmas it is assumed that A, B are points off L such that the line segment AB meets the polygon L in exactly one point, which furthermore is not a vertex of L.

Lemma 5. *One has $\mathcal{I}(A, L) \neq \mathcal{I}(B, L)$.*

Lemma 6. *Every point of the plane off the polygon L can be connected to either A or B by means of a polygonal arc avoiding L.*

We now prove these lemmas. Here is a sketch of the proof of Lemma 3: Let t_1, t_2 be two rays emanating from M. In the case where there are no vertices of L on either of these rays, the intersection of L with either of the sectors determined by the rays consists of simple polygonal arcs with their ends on the rays (somewhat as in the following diagram). From this it is clear that the difference $|t_1 \cap L| - |t_2 \cap L|$ is even, whence $\mathcal{I}(M, L; t_1) = \mathcal{I}(M, L; t_2)$.

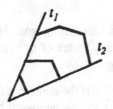

In the case where there *are* vertices of L on the rays, one shows similarly that the difference between the "numbers of points" (in the technical sense defined above in terms of the four possibilities a–d shown in the penultimate diagram) in which L intersects t_1 and t_2 respectively is again an even number; here one needs to examine carefully the aforementioned four possibilities a–d. □

Lemma 4 is proved as follows: Let $A = P_0, P_1, \ldots, P_n = B$ be the vertices of the polygonal arc connecting A and B. Since each segment $P_i P_{i+1}$ avoids L, it follows that the corresponding ray $[P_i P_{i+1})$ and the ray contained in this one, emanating from P_{i+1}, have the same number of points of intersection with L. Hence $\mathcal{I}(A, L) = \mathcal{I}(P_1, L) = \cdots = \mathcal{I}(B, L)$. □

Lemma 5 is proved along similar lines. □

The proof of Lemma 6, which is independent of the others, is based on Theorem 4. Let ℓ be the edge of the polygon Γ containing the point in which the segment AB meets L. Since $L \setminus \ell$ is a simple nonclosed polygonal arc, and hence certainly a tree, there is by Theorem 4 a polygonal arc Γ connecting an arbitrary point $P \notin L$ to A and avoiding $L \setminus \ell$, so that $\Gamma \cap L = \Gamma \cap \ell$. If the latter set is empty, there is nothing to prove. Suppose therefore that there are points of Γ on ℓ, and let M be the first of these encountered on tracing out Γ beginning from P. Write Γ_1 for the portion of Γ from P to M. It is then not difficult to see that any point $C \in \Gamma_1$ sufficiently close to M can be connected either to A or B by means of a polygonal arc avoiding L (see the diagram). $\square$

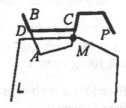

We are now ready to prove the Jordan curve theorem (for simple plane closed polygons): Set

$$U := \{P \in \mathbb{R}^2 \setminus L \mid \mathcal{I}(P, L) = 0\}, \quad V := \{P \in \mathbb{R}^2 \setminus L \mid \mathcal{I}(P, L) = 1\},$$

and let AB be a line segment joining points A, B off the polygon L and meeting L in exactly one point, with that point not a vertex of L. Then by Lemma 5 one of A, B is in U and the other in V, so that, in particular, these sets are not empty. Now let P, Q be any two points of $\mathbb{R}^2 \setminus L$. If one of P, Q is in U and the other in V, then by Lemma 4 they cannot be connected by a polygonal arc avoiding L. On the other hand, if say A, P, $Q \in U$ (in which case necessarily $B \in V$), then (again by Lemma 4) neither P nor Q can be connected to B by means of a polygonal arc avoiding L, while by Lemma 6 they must be able to be so connected to either A or B. Hence both P and Q can be so connected to A, and hence to each other. $\square$

Corollary. *Two plane polygons having no vertices in common intersect in an even number of points.*

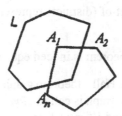

(If L is one of the polygons and $A_1, A_2, \ldots, A_n$ are the vertices of the other, then the number of "jumps" in the sequence $\mathcal{I}(A_1, L), \ldots, \mathcal{I}(A_n, L), \mathcal{I}(A_1, L)$ of zeros and ones, coincides with the number of points of intersection of the two polygons.)

6.6   Pairings

In this section we state and prove a theorem typically exemplified by Problem 3.

Problem 11. (The "marriage problem") [37]. We have a number of young men, each of whom knows several young women. The problem is to find a wife for each young man from among the young women he knows.

Theorem 8 (P. Hall).   *The marriage problem has a solution if and only if for every k, $1 \leq k \leq n$ (where n is the number of young men), every k of the young men know altogether at least k young women.*

The necessity of the stated condition is obvious. We shall prove the sufficiency by induction on n, the number of young men.

Thus suppose inductively that the theorem is true for sets of fewer than n young men, and consider any set of n young men for which the condition of the theorem holds. Suppose, as a first case, that in fact for each k with $1 \leq k \leq n - 1$, every k young men of our set of n know in total at least $k + 1$ young women. By marrying off any young man to any young women he knows, we are then left with $n - 1$ young men, every k of whom know in total at least k young women, for all admissible k, and the inductive assumption applies.

In the contrary case, for some k, $1 \leq k \leq n - 1$, there will be a subset of k young men knowing altogether exactly k young women. Since $k < n$, the inductive assumption applies, so that we can marry off these k young men to young women they know. This leaves $n - k$ young men, each l $(1 \leq l \leq n - k)$ of whom know in total at least l young women (why?). Hence the inductive assumption applies once again, and these $n - k$ young men can also marry appropriately. $\square$

Here is an equivalent abstract formulation of Hall's theorem.

Theorem 9. *Let $\mathcal{E} = (E_1, E_2, \ldots, E_n)$ be an ordered n-tuple of finite sets (not necessarily all distinct). A set $S = \{s_1, s_2, \ldots, s_n\}$ (here the elements are supposed all distinct) with the property that $\forall i$, $s_i \in E_i$, exists if and only if for each k, $1 \leq k \leq n$, the union of any k of the E_i contains at least k elements.*

The set S is called a "set of (distinct) representatives" of the given family of sets $(E_1, \ldots, E_n)$.

Exercise. Show that this theorem is indeed equivalent to Hall's theorem. $\square$

We shall now reformulate Hall's theorem graph-theoretically. We first introduce the appropriate concepts. A graph is called *bipartite* if its vertex–set can be partitioned into two subsets V' and V'' such that every edge of the graph has one of its vertices in V' and the other in V''. A *pairing* from V' to V'' is then a collection $\mathcal{M}$ of edges of the bipartite graph with the property that no two edges share a vertex and every vertex $v' \in V'$ is incident with an edge from $\mathcal{M}$. (In other words, a pairing is an injective map $\varphi : V' \longrightarrow V''$ such that for every vertex $v' \in V'$ there is an edge joining v' to $\varphi(v')$.)

Hall's theorem is then obviously equivalent to the following assertion: In a finite bipartite graph, a pairing from V' to V'' exists if and only if for every subset $A \subseteq V'$ we have $|\varphi(A)| \geq |A|$.

Here are two elementary problems.

Problem 12. On opposite sides of a sheet of paper there are drawn maps in which the same number of countries appear, all of equal area. Prove that one can pierce the sheet of paper with a pin a number of times in such a way that each country (on either side) is pierced exactly once.

An $m \times n$ *Latin rectangle*, $m < n$, is defined to be an $m \times n$ matrix $(a_{ij})_{i=1,\ldots,m}^{j=1,\ldots,n}$ with entries $a_{ij} \in \{1, 2, \ldots, n\}$, such that each row and each column has all its entries distinct.

Problem 13. Prove that any $m \times n$ Latin rectangle can be enlarged to an $n \times n$ Latin square.

(Consider the sets E_j each consisting of those numbers among $1, 2, \ldots, n$ that do not appear as entries in the jth column of the matrix.)

We conclude this section with a theorem which is nontrivial only for infinite graphs.

Theorem 10 (Bernstein–Cantor–Schroeder). *If $\mathcal{M}'$ is a pairing from V' to V'', and $\mathcal{M}''$ a pairing from V'' to V', then there is a pairing $\mathcal{M} \subseteq \mathcal{M}' \cup \mathcal{M}''$ from V' onto V''.*

The proof involves only the subgraph G_0 of the given graph, made up of the edges of the union $\mathcal{M}' \cup \mathcal{M}''$. Since each vertex of this subgraph is incident with either one or two edges, the union of any two simple chains with a vertex in common will again be a simple chain. Hence each vertex of the subgraph G_0 is contained in a maximal (possibly infinite) chain. Those maximal chains consisting of a single edge (joining a vertex in V' to one in V''), if any, will in any case have to be included in $\mathcal{M}$; thus we remove these, together with their endpoints, from our subgraph, leaving a smaller subgraph G_1, say. The maximal chains in G_0 having more than one edge (which are just the maximal chains in G_1) are of the following three possible types:

(a) a closed chain

$$C = (v_0, \ell_0, \bar{v}_0, \bar{\ell}_0, v_1, \ldots, v_0),$$

where $v_i \in V'$, $\ell_i \in \mathcal{M}'$, and $\bar{v}_i \in V''$, $\bar{\ell}_i \in \mathcal{M}''$ (as in Figure a);

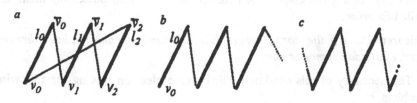

(b) an infinite chain unending on only one side: $C = (v_0, \ell_0, \ldots)$ (Figure b); here v_0 may belong to either V' or V'', and ℓ_0, correspondingly, may be an edge of either M' or M'';

c) an infinite chain unending in both directions: $C = (\ldots, \bar{\ell}_{-1}, v_0, \ell_0, \ldots)$ (Figure c).

For chains of types (a) and (c), the edges of M' determine a pairing between $C \cap V'$ and $C \cap V''$, and so do the edges of M'', so that either will do equally well. In case (b), if the initial vertex v_0 belongs to V', then the edges of M' should be chosen for M, and if $v_0 \in V''$, the edges of M''. □

The Bernstein–Cantor–Schroeder theorem is of fundamental importance for set theory, in particular for "cardinal arithmetic." Recall that two sets A and B are said to have *equal cardinality* if there is a one-to-one correspondence (bijection) between them, and then one writes $|A| = |B|$. If there is a bijection from A onto a subset $B_1 \subseteq B$, then A is said to *have cardinality less than or equal to* that of B, in symbols $|A| \preceq |B|$. In terms of these concepts, the Bernstein–Cantor– Schroeder theorem states that if $|A| \preceq |B|$ and $|B| \preceq |A|$, then $|A| = |B|$.

6.7   Eulerian graphs and a little more

Problem 14. Can one draw the figures  "with one stroke of the pen, i.e., without lifting pencil from paper and without going over any line twice?

Problem 15. (The "Königsberg bridge problem"). Is it possible to stroll around town in such a way as to cross each bridge appearing in the first diagram exactly once?

The latter problem is actually of the same sort as the former, since it is equivalent to that of drawing the graph shown in the second diagram with one stroke of the pen.

In graph-theoretical language, the question becomes: Is there a chain containing every edge of a given graph? Such chains, and the graphs possessing them, are called *Eulerian*.

Theorem 11. *A (finite) connected graph is Eulerian if and only if it has either two vertices of odd valency or none.*

The necessity of this condition will become clear on solving the following problem.

Problem 16. A group of tourists was taken on a guided tour of an archipelago crossing each bridge linking the islands exactly once. It turned out that they visited one of the islands three times. How many bridges must there be to that island if the tour (a) neither started nor ended there? (b) started there but did not end there? (c) both started and ended at that island?

The sufficiency of the condition of the theorem is proved by induction on the number of edges of the graph. The idea involved in the inductive step can be seen by solving the following problem.

Problem 17. Prove that the figure formed by any number of intersecting circles can be drawn with one stroke of the pen. □

Here is one more elementary problem on this theme.

Problem 18. (a) What is the least number of times one needs to cut a piece of wire 120 cm. in length in order to make a wire cube of side 10 cm? (b) What is the shortest length a piece of wire can have for it to be possible to make a wire cube of that size without cutting the wire?

We end this chapter with two problems that are to some extent related to graph-coloring.

Problem 19. (a) Eleven cog-wheels are arranged in a circle, each slotted into the next. Can the cog-wheels of such a system turn? (b) What if there are 12 cog-wheels? (The number of teeth may vary.) (c) Find conditions ensuring that an interconnected system of cog–wheels lying on a plane can move.

Problem 20. In a certain kingdom every two towns are linked either by a highway or by a railway line, but not by both. Prove that one can get from any town to any other by means of bus or train alone, without passing through more than two other towns *en route*.

It can scarcely have escaped notice that the present chapter is not concerned so much with "graph theory" as with graphs as interesting mathematical objects that are exceptionally natural and easy to visualize, notwithstanding that they are of course mathematical abstractions. That the first description applies to them is sufficiently clear; here are two further problems illustrating the second, i.e., their mathematical abstractness:

Problem 21. Does there exist a convex polyhedron with 101 faces, with each face a 3, 5, or 7-gon?

Problem 22. Can one arrange nine straight–line segments on the plane so that each meets exactly three others?

The crucial step in the solutions of both of these problems involves the construction of a certain graph to which Theorem 1 of the present chapter is then applied. In Problem 21 the graph in question is the "dual graph" of the polyhedron. The graph that one is led to investigate in solving Problem 22 also represents a particular case of a general mathematical construction. Note that the latter problem becomes simpler if one replaces the line–segments with arbitrary sets, since this forestalls attempts at a solution using geometric properties of intervals.

As with combinatorial problems, so also may (and should) problems involving graphs be given to junior as well as senior students. In particular, this allows the possibility of introducing to them such important concepts as isomorphism (cf. the simple examples in the book [10]) and mathematical modeling. For the senior students they will also be useful in providing practice in proofs by mathematical induction. Graph-theoretical problems fall very naturally into types (see [7]), and it is for this reason in particular that they are so popular with leaders of mathematical circles (cf. the remarks at the end of Chapter 4). The reader will certainly have noticed that in the present chapter the exposition proceeded on two levels. For instance, to give the rigorous proof of the existence in any finite tree of a "dangling edge" would make sense only for a mathematically rather sophisticated audience.

CHAPTER 7

The Pigeonhole Principle

7.1 Pigeonholes and pigeons

The pigeonhole principle—"if $n+1$ or more pigeons are roosting in n pigeonholes, then some pigeonhole must contain at least two pigeons"—is totally obvious (since on supposing the contrary, one obtains a contradiction immediately by counting the pigeons). It might seem unlikely that such a simple idea could be used to obtain nontrivial results, yet.... We begin, as usual, with some standard elementary problems [10].

Problem 1. In a class of 30 pupils, Alex Johnson made 13 mistakes in writing from dictation, more than anyone else. Prove that there must be at least three pupils who made the same number of mistakes as each other (with of course zero mistakes included as a possibility).

(Here there are 13 "pigeonholes" corresponding to the possible numbers of mistakes from 0 to 12, in which to place 29 "pigeons," i.e., pupils.)

Observe that on the other hand we can say nothing about the number of mistakes such a trio might have made (beyond the datum that it is less than 13), nor whether there were in the whole class no more than three with an equal number of mistakes, or exactly four, or indeed whether or not all 29 pupils made the same number of mistakes.

Problem 2. On a large, flat, white surface black paint is splashed. Show that there must be two points of the surface that are both black or both white and exactly one meter apart.

(Among the vertices of any equilateral triangle of side one meter, at least two must have the same color.)

Problem 3. In a football competition run over several days 18 teams are participating. Prove that on any day during the first round there are at least two teams which have completed the same number of matches.

It might seem that since the number of possiblities for the number of matches completed, namely 0 to 17, is equal to the number of teams, the pigeonhole principle does not apply. However, if there is a team that has played no matches, then no team can have played all 17 of its matches. (The "pigeonholes" corresponding to 0 and 17 cannot both be occupied at the same time.)

Problem 4. Show that of any 51 points in a square of side one, there must be at least three lying inside (a) some square of side $1/5$; (b) some circle of radius $1/7$.

Part (a) is easy. Next, since $2/7 > \sqrt{2}/5$, a circle of radius $1/7$ contains a square of side $1/5$, so that (b) follows from (a).

Problem 5. The 123 residents of a certain apartment building have ages totalling 3813 years. Prove that there are among them 100 whose ages total at least 3100 years.

Suppose that the 100 oldest residents have ages totaling less than 3100; then the youngest of these must be less than 31 years old. Hence the remaining 23 residents must all be younger than 31, so that the total of their ages is certainly less than 713. However, then altogether the residents have ages totaling less than 3813.

This solution represents a particular case of the following general result.

Theorem 1. *Let $a_1, a_2, \ldots, a_n$ be any n real numbers, $n \geq 1$, and write $s := \frac{1}{n} \sum_{i=1}^{n} a_i$, their arithmetic mean. Then for each $k = 1, 2, \ldots, n$ there exist k distinct indices $i_1, i_2, \ldots, i_k$ and k distinct indices $j_1, j_2, \ldots, j_k$ such that $\sum_{l=1}^{k} a_{i_l} \geq ks$ and $\sum_{l=1}^{k} a_{j_l} \leq ks$.*

(Generalize the argument used to solve Problem 5.) □

As the reader may have noticed, learning to solve elementary problems by means of the pigeonhole principle involves also learning to use "proof by contradiction" correctly.

In what follows it is crucial that the pigeonhole principle can be successfully applied not only via an enumeration of the elements of some collection or other.

Problem 6. Certains arcs of a circle are colored red. If the sum of their lengths is less than half of the circumference, show that there must be a pair of diametrically opposite points that are not red.

To see this, observe that if one colors red the arcs symmetric with respect to those originally colored, under reflection in the center of the circle, then firstly, not all of the circle will be colored red, and secondly, the part of the circle still not colored red will be centrally symmetric.

Problem 7. Fifteen napkins of various sizes and shapes are spread on a table, covering it completely. Prove that eight of them can be removed so that the remainder cover at least $7/15$ of the table.

Let $a_1 \geq a_2 \geq \cdots \geq a_{15}$ be the areas of those parts of the napkins in contact with the table. Clearly, $\sum_{i=1}^{7} a_i \geq 7/15$ (see earlier), and if we remove the last eight napkins, the total area of contact of the remaining seven will certainly not decrease.

(Note that on the other hand it need not be the case that there are 7 napkins covering *less* than 7/15 of the table.)

Problem 8. Prove that some integer power of 29 ends in the digits 001.

Since there are exactly a thousand numbers from 0 to 999, there must be among the powers $29^1, 29^2, \ldots, 29^{1001}$ at least two, say 29^k and 29^l, $k > l$, with the same last three digits, whence $29^k - 29^l$ is divisible by 1000. Since $29^k - 29^l = 29^l(29^{k-l} - 1)$ and 29 and 10 are relatively prime, it follows that $29^{k-l} - 1$ is divisible by 1000. Hence the decimal representation of 29^{k-l} ends in 001.

(Note that this also follows from Fermat's "little" theorem—see Section 7.5.)

Problem 9. Show that there is a Fibonacci number ending in three zeros.

There are altogether 10^6 integer pairs (a, b) with $0 \leq a, b \leq 999$, so that among the pairs $\{(a_i, a_{i+1})\}_{i=0}^{10^6}$ of neighboring Fibonacci numbers, there will be at least two, say (a_k, a_{k+1}) and (a_l, a_{l+1}), $k > l$, such that the differences $a_{k+1} - a_{l+1}$ and $a_k - a_l$ are both divisible by 1000. Since by definition of the Fibonacci numbers we have $a_{k-1} = a_{k+1} - a_k$ and $a_{l-1} = a_{l+1} - a_l$, it follows that the difference $a_{k-1} - a_{l-1}$ is also divisible by 1000, and so on. Continuing in this way, we eventually infer that the numbers $a_{k-l+1} - a_1$ and $a_{k-l} - a_0$ are divisible by 1000. Since $a_0 = a_1 = 1$, it follows that $a_{k-l-1} = a_{k-l+1} - a_{k-l}$ is also divisible by 1000, so that its decimal representation ends in three zeros.

7.2   Poincaré's recurrence theorem

Later on in the chapter we shall return to the above sort of elementary problem. In the present section we expound the somewhat different, very general result of the title. Let (X, μ) be a measure space, i.e., a set X, certain subsets A of which are designated as "measurable" (with measure $\mu(A) \geq 0$), and where the measure has the same basic properties as volume, together with that of "countable additivity": given any countable collection of pairwise nonintersecting measurable subsets A_i, $i \in \mathbb{N}$, the union $A := \bigcup_{i=1}^{\infty} A_i$ is measurable and $\mu(A) = \sum_{i=1}^{\infty} \mu(A_i)$.

A map $T : X \longrightarrow X$ is called *measure–preserving* if every measurable set $E \subseteq X$ has measurable complete inverse image, and moreover, $\mu(T^{-1}(E)) = \mu(E)$.

(Although we shall not give the full definition of "measurable set," those among our readers who have never before encountered the term need not feel disquiet: any sets one might normally think of (in an appropriate context) will be measurable.)

Here, then, is Poincaré's recurrence theorem:

Theorem 2 (Poincaré).   *Let (X, μ) be a measure space for which $\mu(X) < \infty$, and let $T : X \longrightarrow X$ be any measure-preserving map. Then for any measurable* *set $E \subseteq X$, there exists for almost all points $x \in E$ a natural number n (depending on x) such that $T^n(x) \in E$.*

Such points are called *recurrent*. The phrase "almost all" signifies that the set of nonrecurrent points has measure zero.

For the proof, consider the set

$$F := E \cap T^{-1}(X \setminus E) \cap T^{-2}(X \setminus E) \cap \cdots .$$

If $x \in F$, then $T^n(x) \notin E$ for any natural number n, and conversely, so that F is precisely the set of nonrecurrent points of E. Hence $T^{-n}(F) \cap F = \emptyset$ for all $n \in \mathbb{N}$. Since

$$T^{-n}(F \cap T^{-k}(F)) = T^{-n}(F) \cap T^{-(n+k)}(F),$$

it follows that the sets $T^{-n}(F)$ are pairwise nonintersecting. On the other hand, since the map T is measure-preserving, we have $\mu(T^{-n}(F)) = \mu F$. Hence in view of the assumption that the measure of the whole space X is finite, we must have $\mu F = 0$. $\square$

Theorem 3. *Let g be the rotation of the circle S^1 through an angle α incommensurable with π (i.e., $\alpha \neq \pi m / n$ for any $m, n \in \mathbb{Z}$). Then the positive orbit of each point $x \in S^1$ (i.e. the set $\{g^n x\}_{n=1}^{\infty}$) is everywhere dense on the circle (i.e. every arc of the circle contains points of that set).*

We use Poincaré's theorem with $X := S^1$ taken to be the unit circle endowed with the usual Lebesgue measure (so that the measure of an arc is just its usual length, in particular, $\mu S^1 = 2\pi$), and $T := g$. Let E be any open arc of the circle. By Poincaré's theorem, for almost every point $P \in E$ there is a natural number n such that $g^n(P) \in E$. Since g is a rotation, so is g^n, and since $g^n(P)$ is back in E, it may be considered as a rotation through an arc shorter than the arc E. It follows that every arc of length at least that of E must contain points of the form $g^{nk}(P)$. The same assertion holds for any point $x \in E$, since there are recurrent points $P \in E$ arbitrarily close to it. $\square$

Exercise. Prove Theorem 3 directly, without appealing to Poincaré's theorem.

Exercise. At which point in the above proof was the assumed incommensurability of α with π used?

Corollary. *For any positive irrational number ξ the set $\Xi := \{k\xi - n \mid k, n \in \mathbb{N}\}$ is everywhere dense in $\mathbb{R}$.*

To see this, consider an (arbitrarily small) interval (a, b) of the real line, let l be a natural number for which $\widetilde{\xi} := l\xi > b$, and g the rotation of the unit circle through the angle $2\pi\widetilde{\xi}$. Write $(\widetilde{a}, \widetilde{b})$ for the arc of the circle forming the image of (a, b) under the map $x \longmapsto (\cos 2\pi x, \sin 2\pi x)$ wrapping the real line round the unit circle. By the preceding theorem, there is a natural number s such that $g^s(1, 0) \in (\widetilde{a}, \widetilde{b})$, which is equivalent to $a < s\widetilde{\xi} - n < b$ for some integer n. By choice of $\widetilde{\xi}$, we must in fact have $n \in \mathbb{N}$. Thus $(sl)\xi - n \in (a, b)$, with $sl, n \in \mathbb{N}$. $\square$

Exercise. Using the irrationality of π, prove that for any $x \in [-1, 1]$ there exists a sequence $\{n_k\}$ of natural numbers such that $\lim_{k \to \infty} \sin n_k = x$.

Problem 10. Prove that a continuous function $f : \mathbb{R} \longrightarrow \mathbb{R}$ satisfying $f(x+1) = f(x)$ and $f(x + \sqrt{2}) = f(x)$ for all $x \in \mathbb{R}$ must be constant.

This may be solved via the following three exercises. A number ω is called a *period* of a function f if $f(x + \omega) = f(x)$ for all $x \in \mathbb{R}$.

Exercise. Prove that the set of periods of a function defined on $\mathbb{R}$ forms a subgroup of $\mathbb{R}$.

Exercise. Prove that if the function is continuous, then its set of periods is closed.

Exercise. Prove that if G is a proper (i.e. $\neq \mathbb{R}$) closed subgroup of $\mathbb{R}$, then there is a number $\alpha \in \mathbb{R}$ such that $G = \{\alpha k \mid k \in \mathbb{Z}\}$.

Here, for completeness, is an alternative solution of Problem 10 avoiding such scholarly terminology. Let c be any real number. By the above corollary, there exist integers a_n, b_n such that $|c - (a_n + b_n\sqrt{2})| < 1/n$. Writing $x_n := a_n + b_n\sqrt{2}$, we have $f(x_n) = f(b_n\sqrt{2}) = f(0)$, and also $f(x_n) \to f(c)$, since $x_n \to c$. Hence $f(c) = f(0)$.

Poincaré's theorem, or more precisely the above corollary, has applications to number theory. The first of the following two problems is ancillary.

Problem 11. Prove that $\lg 2 \ (:= \log_{10} 2)$ is irrational.

Problem 12. Prove that there is a positive power of 2 whose decimal representation begins with 7.

(Use the fact that $2^k = \overline{7 \cdots}$ if and only if there exists a natural number n such that $7 \cdot 10^n \le 2^k < 8 \cdot 10^n$, or $n + \lg 7 \le k \lg 2 < n + \lg 8$.)

7.3 Liouville's theorem

In this section we shall justify the following paradoxical assertion: Suppose that an airtight vessel is separated by means of an impermeable partition into two compartments, one holding a gas and the other a vacuum. If the partition is removed, then it is almost certain that after a sufficient period of time has elapsed the molecules of gas will all have gathered themselves back in the compartment whence they originated (that is, provided that the universe has not disappeared in the meantime...[1]).

Recall that the *Wronskian* $W(t)$ of a an n–tuple $\psi^{(1)}(t), \psi^{(2)}(t), \ldots, \psi^{(n)}(t)$ of solutions of a system $\dot{x} = A(t)x$, $x \in \mathbb{R}^n$, of n ordinary linear differential equations is defined to be the determinant of the matrix whose jth column has as entries the coordinate functions of the jth solution:

$$W(t) := \det\left(\psi_i^{(j)}(t)\right)_{i,j=1,\ldots,n}.$$

It turns out that the Wronskian is a solution of the linear differential equation $\dot{w} = \operatorname{tr} A(t)w$, where $\operatorname{tr} A(t) = \sum_{i=1}^{n} a_{ii}(t)$ is the trace of the matrix of the given system. Therefore,

$$W(t) = \exp\left(\int_{t_o}^{t} \operatorname{tr} A(\tau)d\tau\right) W(t_0),$$

whence the following lemma.

Lemma 1. *If* $\operatorname{tr} A(t) \equiv 0$, *then* $W(t) = \text{const.}$

Consider now an autonomous system of differential equations

$$\dot{x} = f(x), \quad x, f(x) \in \mathbb{R}^n, \quad f \in C^1(\mathbb{R}^n),$$

and let $\varphi(x_0, t)$ be a solution satisfying the initial condition $\varphi(x_0, 0) = x_0$ (i.e., of the corresponding "Cauchy initial value problem"). The theorem on the differentiability of solutions with respect to the initial condition asserts that φ is smooth as a function of the initial value x_0, and further that its Jacobian matrix $\frac{\partial \varphi}{\partial x}$ affords a fundamental system of solutions of the corresponding linear "variational system"

$$\dot{\xi} = \frac{\partial f}{\partial x}(\varphi(x, t))\xi, \quad \xi \in \mathbb{R}^n,$$

where x is regarded as a parameter. In the case where the function φ is sufficiently well–behaved as a function of both x and t, this is easily seen, since then differentiation of the identity $\frac{d}{dt}(\varphi(x, t)) = f(\varphi(x, t))$ with respect to x yields

$$\frac{d}{dt}\left(\frac{\partial \varphi}{\partial x}(x, t)\right) = \frac{\partial f}{\partial x}(\varphi(x, t))\frac{\partial \varphi}{\partial x}(x, t),$$

so that $\frac{\partial \varphi}{\partial x}$ is a solution of the variational system.

Note also that on differentiating the identity $\varphi(x, 0) = x$, we obtain the initial condition

$$\frac{\partial \varphi}{\partial x}(x, 0) = I_n,$$

where I_n denotes the identity $n \times n$ matrix. Thus at $t = 0$ the Wronskian of this solution of the variational system has the value 1.

From our previous discussion of the Wronskian of a solution of a linear system, now applied to the above solution of the variational system, which has matrix $A(t) := \frac{\partial f}{\partial x}(\varphi(x, t))$, yielding $\operatorname{tr} A(t) = \operatorname{tr}\frac{\partial f}{\partial x} = \sum_{i=1}^{n} \frac{\partial f_i}{\partial x_i} = \operatorname{div} f$, we conclude that

$$\det \frac{\partial \varphi(x, t)}{\partial x} = \exp\left(\int_{0}^{t} \operatorname{div} f(\varphi(x, \tau))d\tau\right).$$

Assuming that the solution $\varphi(x, t)$ of the original autonomous system is defined for all $t \in \mathbb{R}$, we define a map $g_t : \mathbb{R}^n \longrightarrow \mathbb{R}^n$ by $g_t(x) := \varphi(x, t)$, representing so to speak an (infinitesimal) motion, measured at time t, along the trajectories of the system through the various points x.

Theorem 4 (Liouville). *If* div $f = 0$, *then the motion along the trajectories of the system* $\dot{x} = f(x)$ *is a measure–preserving transformation (with respect to Lebesgue measure on* $\mathbb{R}^n$*).*

Assume that $U \subset \mathbb{R}^n$ has finite measure. Making the substitution $y = g_t(x)$ in the integral

$$\int_{g_t(U)} dy_1 dy_2 \cdots dy_n = \text{vol}(g_t(U))$$

yields

$$\text{vol}(g_t(U)) = \int_U \det\left(\frac{\partial g_t}{\partial x}\right) dx_1 dx_2 \cdots dx_n$$

$$= \int_U \det \frac{\partial \varphi}{\partial x}(x, t) dx_1 dx_2 \cdots dx_n$$

$$= \int_U dx_1 dx_2 \cdots dx_n = \text{vol}(U),$$

since

$$\det \frac{\partial \varphi(x, t)}{\partial x} = \exp\left(\int_0^t \text{div} f(\varphi(x, \tau)) d\tau\right) = 1. \quad \square$$

We return now to the paradox we started with, concerning molecules of a gas in a container. Denoting by N the number of molecules, the dynamical system they comprise is described by the differential equation

$$\ddot{q} = -\frac{\partial V}{\partial q}, \quad q \in \mathbb{R}^{3N},$$

or, equivalently, by the first-order autonomous system

$$\begin{cases} \dot{q} = p, \\ \dot{p} = -\dfrac{\partial V}{\partial q}, \end{cases}$$

where q is the position vector of the molecules and V is the potential of the system, depending only on the coordinates of the molecules and bounded from below. Since the divergence of the vector field defined by the right-hand side of the latter system is zero (verify!), we have by Liouville's theorem that the motion along its trajectories preserves volume. Now, as we shall see in Chapter 9, a set of the form

$$\{(p, q) \in \mathbb{R}^{6N} \mid \frac{1}{2} \sum_{i=1}^{3N} p_i^2 + V(q) \leq E_0\}$$

is bounded (and so of finite volume), and sent by g_t to itself. Hence by Poincaré's theorem with X taken to be an appropriate such set and E an arbitrarily small neighbourhood of the initial state (p_0, q_0), we see that it is almost certain that the system will eventually enter a state close to the initial one, i.e., we can be almost sure that after a "certain period of time" has elapsed, the molecules will all return to the compartment whence they originated.

7.4   Minkowski's lemma

A *lattice* in $\mathbb{R}^n$ is defined to be a set of the form

$$L = \{ \sum_{i=1}^{n} k_i b_i \mid k_i \in \mathbb{Z} \},$$

where $\{b_i\}_{i=1}^{n}$ is any particular basis for the vector space $\mathbb{R}^n$. The *fundamental region*, or *cell*, of L is then the set

$$P = \{ \sum_{i=1}^{n} \xi_i b_i \mid \xi \in [0, 1) \},$$

with volume $\mathrm{vol}(P) = |\det(b_i)|$.

Lemma 2. *Given any point $x \in \mathbb{R}^n$, there is exactly one point $y \in P$ such that $x - y \in L$.*

Clearly, the point x with coordinates defined by $x_i \in [0, 1)$, $y_i \equiv x_i \pmod{1}$ is such a point. Its uniqueness follows from the fact that $P \cap L = \{0\}$. $\square$
We thus obtain a map $p : \mathbb{R}^n \longrightarrow P$.

Lemma 3. *The map p is locally volume-preserving; more precisely, if $E \subseteq \mathbb{R}^n$ is a measurable set such that $p(x) \neq p(y)$ for every two distinct points $x, y \in E$, then $\mathrm{vol}\, p(E) = \mathrm{vol}\, E$.*

To see this, first express E as the following disjoint union:

$$E = \bigcup_{u \in L} (E \cap P_u) = \bigcup_{u \in L} E_u,$$

where for each $u \in L$ the set P_u is the translate of the cell P through the vector u; hence the P_u partition $\mathbb{R}^n$, and if $x \in P_u$, then $p(x) = x - u$. From the injectivity of the restriction of the map p to E, it follows that the image sets $p(E_u)$ are pairwise nonintersecting, whence

$$\mathrm{vol}\, p(E) = \sum_{u \in L} \mathrm{vol}\, p(E_u) = \sum_{u \in L} \mathrm{vol}\, (E_u) = \mathrm{vol}\, E. \square$$

By way of illustrating this proof, consider the integer lattice in the plane, i.e., the lattice of points with integer coordinates, and for E take the disk of radius $\frac{1}{3}$ and center $(1, 2)$. The image of this disk under p consists of four quarter-disks of the same radius $\frac{1}{3}$, arranged in the unit square as shown in the following diagram.

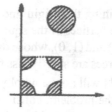

Theorem 5 (Minkowski's lemma [25]).   *Let K be a convex subset of $\mathbb{R}^n$, centrally symmetric with respect to the origin as center, and let L be an arbitrary lattice in $\mathbb{R}^n$, with fundamental region P. If $\mathrm{vol}\,K > 2^n \mathrm{vol}\,P$, then the set K contains at least one point of the lattice apart from the origin.*

Consider the set K' homothetic to K, with dilation factor $\frac{1}{2}$. Then $\mathrm{vol}\,K' = 2^{-n}\mathrm{vol}\,K > \mathrm{vol}\,P$, so that since the map p is locally volume-preserving, its restriction $p|_{K'} : K' \longrightarrow P$ cannot be injective. Hence there exist distinct points $x, y \in K'$ such that $z := x - y \in L$. Then since $2x, 2y \in K$, and K is convex and centrally symmetric about the origin, we have $z = (2x + (-2y))/2 \in K$. $\square$

Corollary.   *Any lattice $L \subset \mathbb{R}^n$ with fundamental region of volume v contains a point $\mathbf{x} \neq O$ whose distance from the origin is at most $2\sqrt[n]{v/\omega_n}$, where ω_n denotes the volume of the n-dimensional unit ball.*

Denote by B the closed n-ball of radius r, centered at the origin. Since B is closed, it follows from Minkowski's lemma that if $\mathrm{vol}\,B \geq 2^n v$, i.e., if $r^n \omega_n \geq 2^n v$, then B contains a nonzero point of L. Since the least value of r ensuring this is $2\sqrt[n]{v/\omega_n}$, there must be a point of L ($\neq O$) not further than this from the origin. $\square$

In the succeeding two sections we shall use Minkowski's lemma to help prove two classical results of number theory. In the meantime, we shall in the present section use it to solve some rather more elementary problems.

Problem 13.   Consider the set of disks of some fixed radius with their centers at the vertices of a tesselation of the plane into squares of fixed side. Show that any straight line passing through a vertex of the tesselation must meet at least one other disk.

(Use the corollary to Theorem 3.)

Thus in an infinite plantation of trees planted so as to form a regular square array, no matter how slender the trees are, from a vantage point adjacent to any tree one will see nothing but tree–trunks in every direction.

Problem 14 ([38]).   Imagine a garden in the form of a disk of radius 50 meters, planted with trees in a square array of side one meter. At the center, however, in place of a tree there is a summer–house. Show that as long as the radii of the trees remain less than $\frac{1}{\sqrt{2501}}$, the view from the summer–house will not be entirely blocked, but when they exceed $\frac{1}{50}$ there will be nothing but tree–trunks visible.

Consider the straight line joining the origin and the point $P(50, 1)$ (which is outside the garden). Clearly, the vertices of the square array inside the garden that are closest to this line are $(49, 1)$ and $(1, 0)$, whose distances from the line are both $\frac{1}{\sqrt{2501}}$. Hence if the radii of the trees are strictly less than $\frac{1}{\sqrt{2501}}$, then the view from the center along this line of sight will not be wholly impeded by the trees.

To solve the second half of the problem, consider an arbitrary diameter of the garden, and let $ABCD$ be the rectangle of length 100 and width 1/25, having that diameter through its middle (so that $BC = 100$ and $AB = 1/25$, as in the diagram.)

Since the fundamental region P of the given lattice is the unit square, and the area of the rectangle $ABCD$ is $100/25 = 4$, we have that that area is equal to $2^2 \text{vol } P$. Hence in view of the convexity and central symmetry of the rectangle, Minkowski's lemma guarantees that it contains a lattice point other than the origin, which must then of course be at most 1/50 of a meter from the chosen diameter. Thus if the radii of the tree–trunks exceed 1/50, then the view from the center in the direction of any diameter will encounter a trunk.

Here is another application of Minkowski's lemma, this one concerning approximation of real numbers by rationals [15].

Problem 15. Prove that for any real number α there is a rational number m/n with arbitrarily large denominator n such that $|\alpha - m/n| \leq 1/n^2$.

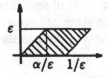

The case where α is rational being trivial, we assume α irrational. Clearly, we may also suppose that $\alpha \in (0, 1)$. Consider the plane lattice

$$\{(x, y) \mid x = \frac{\alpha n - m}{\varepsilon}, y = \varepsilon n, \ m, n \in \mathbb{Z}\},$$

where ε is any fixed positive real number. The fundamental region of this lattice (see the diagram) has area 1. Since the square with vertices $(\pm 1, \pm 1), (\pm 1, \mp 1)$ fulfils all of the conditions of Minkowski's lemma with respect to this lattice, it must contain a lattice point other than the origin. Hence there exist integers m, n

satisfying $|(\alpha n - m)/\varepsilon| \le 1$ and $|\varepsilon n| \le 1$, whence $|\alpha - m/n| \le \varepsilon/|n| \le 1/n^2$, and by taking ε sufficiently small we can force n to be as large as we please.

7.5 Sums of two squares

The goal in this section and the next is to prove the following two theorems.

Theorem 6 (Fermat–Euler). *An odd prime number p can be expressed as the sum of two squares of integers, $p = a^2 + b^2$, $a, b \in \mathbb{Z}$, if and only if $p \equiv 1 \pmod 4$.*

Theorem 7 (Lagrange). *Every natural number is the sum of four squares of integers.*

The present section is devoted chiefly to proving the first of these theorems. The necessity is relatively easy:

Exercise. Show that any odd natural number expressible as a sum of two squares must be of the form $4k + 1$.

In the initial part of the proof in the other—more significant—direction, we shall work with the ring $\mathbb{Z}_n = \{0, 1, \ldots, n - 1\}$ of residues modulo n (i.e., remainders after division by n). We denote by $\mathbb{Z}_n^*$ the subset of *units*, i.e., invertible elements with respect to multiplication:

$$\mathbb{Z}_n^* := \{a \in \mathbb{Z}_n \mid \exists b \in \mathbb{Z}_n : ab \equiv 1 \pmod n\}.$$

Lemma 4. *The set $\mathbb{Z}_n^*$ consists of the natural numbers less than and relatively prime to n.*

For $ab \equiv 1 \pmod n$ means that $ab = 1 - mn$ for some $m \in \mathbb{Z}$, i.e., $ab + mn = 1$, which is equivalent to the coprimality of a and n. $\square$

At this point we introduce a basic result of the theory of groups, namely "Lagrange's theorem": If G is a finite group and H a subgroup, then the order $|H|$ of H (i.e., the number of elements of H) divides the order $|G|$ of G. This provides one avenue to proving the following well-known theorem.

Theorem 8 (Fermat's "little" theorem). *If p is a prime and t any integer not divisible by p, then $t^{p-1} \equiv 1 \pmod p$.*

Note first that since p is prime, we have $|\mathbb{Z}_p^*| = p - 1$. Let $1, t, \ldots, t^{k-1}$ be all of the powers of t that are distinct modulo p. After being reduced modulo p, these form a subgroup of order k of the multiplicative group $\mathbb{Z}_p^*$. Hence by Lagrange's theorem k must divide $p - 1$, whence $t^{p-1} = (t^k)^{(p-1)/k} \equiv 1 \pmod p$.

Here is a proof—the standard one—not using group theory. Since $t \not\equiv 0 \pmod p$, the numbers kt, $k = 1, 2, \ldots, p - 1$, must have distinct nonzero residues modulo p. Hence $(p - 1)! t^{p-1} \equiv (p - 1)! \pmod p$, which yields $t^{p-1} \equiv 1 \pmod p$. $\square$

Exercise. (Euler). Let n be any natural number ≥ 2, and a any integer relatively prime to n. Prove that $a^{\varphi(n)} \equiv 1(\bmod\ n)$, where $\varphi(n)$ denotes the number of natural numbers less than n and relatively prime to n.

For the next—and last—lemma on our way to proving the Euler-Fermat theorem, we shall need a special case of a result proved earlier (Theorem 10 of Chapter 5), namely that for every prime p the (multiplicative) group $\mathbb{Z}_p^*$ is cyclic, i.e., there exists an element $t \in \mathbb{Z}_p^*$ with the property that $\mathbb{Z}_p^* = \{1, t, \ldots, t^{p-2}\}$.

Exercise. Show that the group $\mathbb{Z}_{15}^*$ is not cyclic.

Lemma 5. *Let p be an odd prime. Then the equation $x^2 + 1 = 0$ has a solution in $\mathbb{Z}_p$ if and only if $p \equiv 1(\bmod\ 4)$.*

Necessity: Suppose $u \in \mathbb{Z}_p$ is a solution, i.e., $u^2 + 1 = 0$. Then $u^4 = 1$, so that the set $\{1, u, u^2, u^3\}$ constitutes a subgroup of $\mathbb{Z}_p^*$. Hence by Lagrange's theorem, 4 must divide $p - 1$.

Sufficiency: Suppose that 4 divides $p - 1$. Let t be a generator of $\mathbb{Z}_p^*$, and write $u := t^{(p-1)/4}$ and $y := u^2 = t^{(p-1)/2}$. Then since $y^2 = 1$ in $\mathbb{Z}_p$, we have $(y - 1)(y + 1) = 0$. Since p is prime, there are no zero–divisors in $\mathbb{Z}_p$ (it is in fact a field), whence either $y = 1$ or $y = -1$ in $\mathbb{Z}_p$. The equation $y = 1$ translates into $t^{(p-1)/2} = 1$, which contradicts the choice of t as a generator of $\mathbb{Z}_p^*$. Hence we must have $y = -1$. $\square$

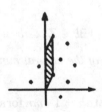

We are now ready to complete the proof of Fermat's theorem. Thus let p be a prime of the form $4k + 1$. Then by Lemma 5 there exists an element $u \in \mathbb{Z}_p$ such that $u^2 \equiv -1(\bmod\ p)$. Consider the plane lattice

$$L := \{(a, b) \mid a, b \in \mathbb{Z},\ b \equiv ua(\bmod\ p)\}.$$

(The diagram depicts this lattice for $p = 5$.) As generating vectors for this lattice we may take $\mathbf{b}_1(1, u)$ and $\mathbf{b}_2(0, p)$. Hence the area of the fundamental region is just p. Since the area ω_2 of the unit circle is π, we infer from the corollary to Minkowski's lemma that there exists a nonzero vector $\mathbf{x} = (a, b) \in L$ satisfying $|\mathbf{x}|^2 = a^2 + b^2 \leq 4p/\pi < 2p$. On the other hand, we have

$$a^2 + b^2 \equiv a^2 + u^2a^2 \equiv a^2(1 + u^2) \equiv 0(\bmod\ p),$$

i.e., p divides $a^2 + b^2$. Hence $a^2 + b^2 = p$. $\square$

Corollary. *Let n be any natural number, and $n = p_1^{k_1} p_2^{k_2} \cdots p_s^{k_s}$ its decomposition as a product of positive powers of distinct primes p_i. Then n can be written as* the sum of two squares of integers if and only if k_i is even for those p_i satisfying $p_i \equiv -1 \pmod 4$.

The sufficiency follows easily from the following two implications: $n = a^2 + b^2 \implies k^2 n = (ka)^2 + (kb)^2$ and $(n = a^2 + b^2) \wedge (m = c^2 + d^2) \implies mn = (ac + bd)^2 + (ad - bc)^2$.

The necessity can be established as follows. Suppose that $n = a^2 + b^2$ for some $a, b \in \mathbb{Z}$, and that in the prime decomposition $n = p_1^{k_1} p_2^{k_2} \cdots p_s^{k_s}$ we have $p_1 \equiv -1 \pmod 4$, and k_1 odd, say $k_1 = 2l + 1$. Let p_1^{2m} be the largest power of p_1 dividing both a^2 and b^2. Then $l \geq m$, and

$$\left(\frac{a}{p_1^m}\right)^2 + \left(\frac{b}{p_1^m}\right)^2 = p_1^{1+2(l-m)} p_2^{k_2} \cdots p_s^{k_s} \equiv 0 \pmod{p_1},$$

which we rewrite more concisely as $(a')^2 + (b')^2 \equiv 0 \pmod{p_1}$. By definition of m, at least one of the integers a', b' is not divisible by p_1; without loss of generality we may suppose that $a' \not\equiv 0 \pmod{p_1}$. Then since $\mathbb{Z}_{p_1}$ is a field, there exists an integer a_1 such that $a' a_1 \equiv 1 \pmod{p_1}$. However then $(b' a_1)^2 \equiv -(a' a_1)^2 \equiv -1 \pmod{p_1}$, which by Lemma 5 is impossible. $\square$

7.6 Sums of four squares. Euler's identity

For the proof we give here of Lagrange's theorem, we shall need two lemmas.

Lemma 6. *The volume ω_4 of the unit 4-dimensional ball is $\pi^2/2$.*

The unit 4-dimensional ball is defined by $B^4 := \{(x_1, x_2, x_3, x_4) \mid \sum_{i=1}^4 x_i^2 \leq 1\}$. Write D^2 for the 2-dimensional ball (i.e., disk) in which B^4 intersects the plane Ox_1x_2 determined by the x_1- and x_2-axes. For each point $(x_1, x_2, 0, 0)$ of this disk, consider the intersection of the ball B^4 with the plane parallel to the plane Ox_3x_4 and passing through that point; this intersection is a disk $D_r^2(x_1, x_2)$ of radius $r = \sqrt{1 - x_1^2 - x_2^2}$. By Fubini's theorem, allowing multiple integrals to be evaluated by means of appropriate iterated integrals (in other words, as an integral of the areas of cross-sections), we have

$$\omega_4 = \operatorname{vol} B^4 = \int_{B^4} dx_1 dx_2 dx_3 dx_4$$

$$= \int_{D^2} \left(\int_{D_r^2(x_1,x_2)} dx_3 dx_4 \right) dx_1 dx_2$$

$$= \int_{D^2} \pi r^2 dx_1 dx_2 = \pi \int_{D^2} (1 - x_1^2 - x_2^2) dx_1 dx_2$$

$$= \pi \int_0^{2\pi} \left(\int_0^1 (1 - \rho^2) \rho d\rho \right) d\varphi = 2\pi^2 \left(\frac{\rho^2}{2} - \frac{\rho^4}{4} \right) \Big|_0^1 = \frac{\pi^2}{2},$$

where we have used the standard substitution $x_1 = \rho \cos \varphi$, $x_2 = \rho \sin \varphi$ of polar for plane rectangular coordinates, which has Jacobian

$$\begin{vmatrix} \frac{\partial x_1}{\partial \rho} & \frac{\partial x_2}{\partial \rho} \\ \frac{\partial x_1}{\partial \varphi} & \frac{\partial x_2}{\partial \varphi} \end{vmatrix} = \begin{vmatrix} \cos \varphi & \sin \varphi \\ -\rho \sin \varphi & \rho \cos \varphi \end{vmatrix} = \rho. \quad \square$$

Lemma 7. *For every prime p, the equation $x^2 + y^2 + 1 = 0$ has a solution in $\mathbb{Z}_p$.*

The case $p = 2$ being obvious, we may assume that p is odd. Consider the two maps $f, g : \mathbb{Z}_p \to \mathbb{Z}_p$, defined by $f(u) := u^2$ and $g(u) := -1 - u^2$. Clearly, the image sets $f(\mathbb{Z}_p)$ and $g(\mathbb{Z}_p)$ have the same size. The image set under f consists of zero together with one element u^2 for each pair $-u, u \in \mathbb{Z}_p$, $u \neq 0$, making altogether $1 + (p-1)/2 = (p+1)/2$ elements. Since $(p+1)/2 + (p+1)/2 > p$, by the pigeonhole principle the image sets of f and g must intersect, i.e., there are elements $u, v \in \mathbb{Z}_p$ satisfying $u^2 = -1 - v^2$, whence $u^2 + v^2 + 1 = 0$. $\square$

Proceeding to the proof of Lagrange's theorem, assume to begin with that $n = p$, a prime. Then by Lemma 7 there exist elements $u, v \in \mathbb{Z}_p$ such that $u^2 + v^2 + 1 = 0$. Consider the lattice $L \subset \mathbb{Z}^4$ in $\mathbb{R}^4$, defined by

$$L := \{(a, b, c, d) \mid c \equiv ua + vb \pmod{p}, \ d \equiv ub - va \pmod{p}\}.$$

Exercise. Show that the vectors $\mathbf{b}_1(1, 0, u, -v)$, $\mathbf{b}_2(0, 1, v, u)$, $\mathbf{b}_3(0, 0, p, 0)$ and $\mathbf{b}_4(0, 0, 0, p)$, form a basis for the lattice L, and that the volume of the fundamental domain is p^2.

Hence by the corollary to Minkowski's lemma, the lattice L contains a nonzero point $\mathbf{x} = (a, b, c, d)$ such that

$$|\mathbf{x}|^2 = a^2 + b^2 + c^2 + d^2 \le 4\sqrt{\frac{p^2}{\omega_4}} = 4\sqrt{\frac{2p^2}{\pi^2}} = p\frac{4\sqrt{2}}{\pi} < 2p.$$

On the other hand, since

$$\begin{aligned} |\mathbf{x}|^2 &= a^2 + b^2 + (ua + vb)^2 + (ub - va)^2 \\ &= (a^2 + b^2)(1 + u^2 + v^2) \equiv 0 \pmod{p}, \end{aligned}$$

we have that $|\mathbf{x}|^2$ is divisible by p. From this and the inequality $|\mathbf{x}|^2 < 2p$, we infer immediately that $|\mathbf{x}|^2 = a^2 + b^2 + c^2 + d^2 = p$.

The conclusion for arbitrary natural numbers n now follows from Euler's identity

$$\begin{aligned} (a^2 + b^2 &+ c^2 + d^2)(A^2 + B^2 + C^2 + D^2) \\ &= (aA - bB - cC - dD)^2 + (aB + bA + cD - dC)^2 \\ &+ (aC + cA + bD - dB)^2 + (aD + dA + bC - cB)^2, \end{aligned}$$

since this shows that the product of two sums of four squares of integers is again a sum of four squares of integers.

Exercise. Verify Euler's identity.

We conclude this chapter with the following observation (for the reader to conjure with): if $L \subseteq L' \subset \mathbb{R}^n$, where L and L' are lattices with fundamental regions

P, P' of volumes v, v' respectively, then $v = |L'/L|v'$, where L'/L denotes the quotient group of L' by L, regarded as additive subgroups of $\mathbb{R}^n$. This follows from properties of the map $p : P \to P'$, namely that it is, firstly, locally volume-preserving, and secondly, $|L'/L|$-fold; i.e., each point $x \in P'$ is the image of exactly $|L'/L|$ elements of P.

CHAPTER 8

The Quaternions

8.1 The skew–field of quaternions, and Euler's identity

This chapter contains no elementary problems; its inclusion was prompted chiefly by internal necessity. The main motive was a desire to elucidate the strange and beautiful identity introduced at the end of the preceding chapter, known as Euler's identity. Furthermore, the skew–field of quaternions is an archetype of a general (and important) kind of algebraic structure, namely that of an "algebra," and so serves as a "jumping-off point" for studying such objects. Lastly, we have as a consequence of one of the main theorems of this chapter that the chain of number–fields $\mathbb{Q} \subset \mathbb{R} \subset \mathbb{C}$ extends no further.

Consider the vector space $\mathbb{R}^4$ with the usual (componentwise) addition (ignoring the scalar multiplication for the time being), with respect to which it is an abelian group. Let us try to define an appropriate multiplication (i.e., a product $u \cdot v$ for each pair $u, v \in \mathbb{R}^4$, which, in conjunction with addition, obeys the basic rules of arithmetic). Componentwise multiplication (defined by $u \cdot v := (u_0 v_0, u_1 v_1, u_2 v_2, u_3 v_3)$, where $u = (u_0, u_1, u_2, u_3)$, $v = (v_0, v_1, v_2, v_3)$) will not do, since with respect to this multiplication there are zero–divisors, i.e., nonzero elements $u, v \in \mathbb{R}^4$ such that $u \cdot v = 0$, and this rules out division by such elements. The product defined in terms of matrix multiplication by

$$\begin{pmatrix} u_0 & u_1 \\ u_2 & u_3 \end{pmatrix} \begin{pmatrix} v_0 & v_1 \\ v_2 & v_3 \end{pmatrix} = \begin{pmatrix} u_0 v_0 + u_1 v_2 & u_0 v_1 + u_1 v_3 \\ u_2 v_0 + u_3 v_2 & u_2 v_1 + u_3 v_3 \end{pmatrix}$$

has the same defect.

Let e, i, j, k denote the standard basis for $\mathbb{R}^4$, and define the various products of these elements by:

$$i^2 = j^2 = k^2 = -e^2 = -e,$$

$$ei = ie = i, \quad ej = je = j, \quad ek = ke = k,$$
$$ij = -ji = k, \quad jk = -kj = i, \quad ki = -ik = j.$$

Exercise. Verify that the product of any three of these elements obeys the associative law.

(One simply checks the 27 nonobvious equations systematically. Later on, associativity will be established by other means.)

Lemma 1. *There is precisely one map $f : \mathbb{R}^4 \times \mathbb{R}^4 \to \mathbb{R}^4$ that is linear in each variable (bilinear), and whose value at each ordered pair of the basis elements e, i, j, k is the product of that pair. Moreover, that unique map has the property that whenever $f(u, v) = 0$, one has either $u = 0$ or $v = 0$.*

(Recall that the expression *linear in each variable*, or *multilinear* (in the present case *bilinear*, since there are just two variables), means here that for any elements $u, v, w \in \mathbb{R}^4$ and numbers $\alpha, \beta \in \mathbb{R}$ one has $f(\alpha u + \beta v, w) = \alpha f(u, w) + \beta f(v, w)$ and $f(u, \alpha v + \beta w) = \alpha f(u, v) + \beta f(u, w)$.)

The existence and uniqueness of such a bilinear function is easily verified. Here is the proof of the final assertion of the theorem. For any elements $u = ae + bi + cj + dk$, $v = Ae + Bi + Cj + Dk$ of $\mathbb{R}^4$, we have, using the bilinearity of f and its prescribed value at each pair of basis elements (e.g. $f(i, j) := ij = k$), that

$$f(u, v) = (aA - bB - cC - dD)e + (aB + bA + cD - dC)i$$
$$+ (aC + cA + dB - bD)j + (aD + dA + bC - cB)k.$$

Suppose now that $f(u, v) = 0$. If $u = 0$, there is nothing to prove, so we assume that $u \neq 0$, and proceed to show that we must then have $v = 0$. In view of the above formula for $f(u, v)$, the components A, B, C, D of v represent a solution of the system

$$ax - by - cz - dt = 0,$$
$$bx + ay - dz + ct = 0,$$
$$cx + dy + az - bt = 0,$$
$$dx - cy + bz + at = 0.$$

It is easy to check that the column vectors of the coefficient matrix of this system are pairwise orthogonal and have (Euclidean) length $\sqrt{a^2 + b^2 + c^2 + d^2} \neq 0$. Now, the determinant of a square matrix over $\mathbb{R}$ is, in absolute value, the volume of the parallelepiped determined by its column (or row) vectors. In the present case this is a 4–dimensional cube, so that certainly the determinant of the above homogeneous system is nonzero. Hence that system has only the trivial solution, whence we must have $v = 0$. $\square$

Thus we take our product operation in $\mathbb{R}^4$ to be defined by $u \cdot v := f(u, v)$, as given explicitly by the above expression for $f(u, v)$.

Given any element $u = ae + bi + cj + dk$ of this system with its two operations, we call the element $\bar{u} := ae - bi - cj - dk$ the *conjugate* of u. The following properties of conjugation are easily verified.

Lemma 2. *One has the following identities:*
(a) $u \cdot \bar{u} = \bar{u} \cdot u = a^2 + b^2 + c^2 + d^2 =: |u|^2$.
(b) $u \cdot (\bar{u}/|u|^2) = e$ *(provided $u \neq 0$)*.
(c) $\overline{u \cdot v} = \bar{v} \cdot \bar{u}$. $\square$

Theorem 1. *The set $\mathbb{R}^4$ equipped with the above multiplication and the usual (vector) addition is a skew–field (i.e., satisfies all of the defining conditions of a field, except for commutativity of multiplication).*

The elements of this skew–field are called *quaternions*. It is customary to denote it by $\mathbb{H}$, in honor of its discoverer, W.R. Hamilton.

The distributivity of multiplication over addition follows from the bilinearity of the map f. The multiplicative identity element is e, since it was postulated that $ei = ie = i$, $ej = je = j$, $ek = ke = k$, and $ee = e$, whence, again by the bilinearity,

$$e(ae + bi + cj + dk) = a(ee) + b(ei) + c(ej) + d(ek)$$
$$= ae + bi + cj + dk,$$

and similarly for $(ae + bi + cj + dk)e$. From the identities (a) and (b) of Lemma 2, we see that the multiplicative inverse of any element $u \neq 0$ is $\bar{u}/|u|^2$. The associative law of multiplication follows from its validity for the basis elements, and, once again, the bilinearity of f. $\square$

Exercise. Prove that the subset $\{ae \mid a \in \mathbb{R}\}$ is a subfield of $\mathbb{H}$ isomorphic to $\mathbb{R}$.

(Use $(ae)(be) = (ab)e$.)

From now on we shall omit the symbol e from the notation for quaternions (or replace it by 1); thus a quaternion $u = (a, b, c, d) \in \mathbb{R}^4$ will henceforth be written in the form $u = a + bi + cj + dk$. The quantity $|u| := \sqrt{a^2 + b^2 + c^2 + d^2}$ is called the *norm* of the quaternion u.

Theorem 2. *In $\mathbb{H}$ one has the identity $|uv| = |u||v|$ for all quaternions u, v (equivalent to Euler's identity).*

By Lemma 2, we have

$$|uv|^2 = (uv)(\overline{uv}) = (uv)(\bar{v}\bar{u}) = (u(v\bar{v}))\bar{u}$$
$$= (u|v|^2)\bar{u} = |v|^2(u\bar{u}) = |u|^2|v|^2.$$

On expanding this identity in terms of components, using the formula for $f(u, v) = u \cdot v$ given earlier, one obtains Euler's identity. $\square$

8.2   Division algebras. Frobenius's theorem

An *algebra* over a field F is a vector space V over F on which there is defined a multiplication $u \cdot v$, $u, v \in V$, determining a bilinear map $V \times V \longrightarrow V$ $((u, v) \longmapsto u \cdot v)$.

Theorem 3 (Stiefel [16]).   *If a product operation can be defined on the space* $\mathbb{R}^n$ *so as to yield an $\mathbb{R}$-algebra without zero–divisors, then on the sphere S^{n-1} one can define $n - 1$ continuous, (pointwise) linearly independent, tangent vector fields.*

Let $b_1, b_2, \ldots, b_n$ form a basis for $\mathbb{R}^n$, and for each $i = 1, \ldots, n$, consider the map $\varphi_i : \mathbb{R}^n \to \mathbb{R}^n$ defined in terms of the multiplication by $\varphi_i(x) := x \cdot b_i$. By the bilinearity of multiplication, each map φ_i is linear, moreover, since there are no zero–divisors, with trivial kernel. Hence the φ_i are vector–space isomorphisms, whence so are the maps $v_i : \mathbb{R}^n \to \mathbb{R}^n$ defined for each i by $v_i := \varphi_i \varphi_1^{-1}$. Note that v_1 is the identity map. The φ_i have the property that for every nonzero vector $x \in \mathbb{R}^n$, the images $\varphi_1(x), \varphi_2(x), \ldots, \varphi_n(x)$ are linearly independent, and so form a basis for $\mathbb{R}^n$; for, $\sum_{i=1}^n \lambda_i \varphi_i(x) = 0$ is equivalent to $\sum_{i=1}^n \lambda_i(x b_i) = x \sum_{i=1}^n \lambda_i b_i = 0$, whence $\sum_{i=1}^n \lambda_i b_i = 0$, which, since the b_i are linearly independent, implies that the λ_i are all zero. Hence by their definition in terms of the φ_i, the maps v_i have the same property. Thus if we restrict the v_i to the unit sphere $S^{n-1} \subset \mathbb{R}^n$, we obtain n linearly independent vector fields on S^{n-1}. In view of the bilinearity, the maps φ_i are certainly continuous, whence so are these vector fields. From these we obtain, for each $i = 2, 3, \ldots, n$, a vector field $\hat{v}_i$ *tangential* to S^{n-1}, taking $\hat{v}_i(y)$, for each vector y with its tip on S^{n-1}, to be the projection of $v_i(y)$ onto the hyperplane through O perpendicular to y. Thus for each $i = 2, 3, \ldots, n$, $\hat{v}_i(y)$ has the form $\hat{v}_i(y) = v_i(y) + v_i y$ for some real number v_i. Since $v_1(y) = y$, and $v_1(y), v_2(y), \ldots, v_n(y)$ are linearly independent, the $n - 1$ tangent vector fields $\hat{v}_2, \hat{v}_3, \ldots, \hat{v}_n$ must also be linearly independent. $\square$

There is a difficult theorem of topology asserting that the only spheres with this property are S^1, S^3, and S^7. It is considerably less difficult, though still not elementary, to establish the particular case of this result, that there is no continuous nonvanishing tangent vector field on S^2 (sometimes called "the uncombability of a hedgehog").

Exercise. Find in explicit form three continuous linearly independent tangent vector fields on S^3.

A *division algebra* over a field F is defined to be an associative F–algebra with a multiplicative identity element in which every nonzero element has a multiplicative inverse (in other words, a skew–field extending F). We shall need the following three lemmas; their proofs will largely be left to the reader as exercises.

Lemma 3. *In a finite-dimensional F-algebra A with an identity element, every element α is "algebraic" over F; i.e., there exist elements $a_0, a_1, \ldots, a_N \in F$, not all zero, such that $\sum_{i=0}^N a_i \alpha^i = 0$.*

(The $N + 1$ elements $1, \alpha, \ldots, \alpha^N$ must be linearly dependent, where $N :=$ $\dim_F A$.) □

Lemma 4. *Let A be an F-algebra with an identity element. If $f, g \in F[t]$ are polynomials with coefficients from F, and $h := fg$, then for every element $\alpha \in A$, one has $h(\alpha) = f(\alpha)g(\alpha)$.* □

Lemma 5. *If A is an F–algebra with an identity element, then $F \subseteq A$; more precisely, A contains a subalgebra isomorphic to F.*

(See the exercise following Theorem 1.) □

Theorem 4. (a) *Every finite-dimensional, commutative division algebra A over $\mathbb{R}$ is isomorphic to either $\mathbb{R}$ or $\mathbb{C}$.*

(b) *Every division algebra over $\mathbb{R}$ of dimension 1 or 2 is isomorphic to $\mathbb{R}$ or $\mathbb{C}$.*

Here is a proof of the first assertion. By Lemma 5, we have $\mathbb{R} \subseteq A$. In the case of equality there is nothing to prove, so we may suppose that there is an element $\alpha \in A \setminus \mathbb{R}$. By Lemma 3, there is a nonconstant polynomial $f \in \mathbb{R}[t]$ such that $f(\alpha) = 0$. By the fundamental theorem of algebra, f factors as a product of linear or quadratic polynomials over $\mathbb{R}$, say $f(t) = f_1(t)f_2(t)\cdots f_k(t)$, where $f_j \in \mathbb{R}[t]$ and $\deg f_j = 1$ or 2. Then by Lemma 4, we have $f(\alpha) = f_1(\alpha)\cdots f_k(\alpha)$, whence we infer, since A has no zero–divisors, that $f_j(\alpha) = 0$ for some j. The degree of this f_j must be 2, since otherwise we should have $\alpha \in \mathbb{R}$, and f_j may be assumed monic by dividing out its leading coefficient; write $f_j = t^2 + bt + c$. The discriminant of this quadratic polynomial must be negative, i.e., $b^2 - 4c < 0$, since otherwise we should again have $\alpha \in \mathbb{R}$. let d be a real number satisfying $-d^2 = b^2/4 - c$; then $(\alpha/d + b/(2d))^2 = -1$. Write $i := \alpha/d + b/(2d)$. Since $\alpha \notin \mathbb{R}$, the elements 1 and i are certainly linearly independent; denote by A_0 the subalgebra they generate (which is the same as that generated by 1 and α). Clearly, A_0 is isomorphic to $\mathbb{C}$.

We shall now show that in fact $A_0 = A$. Let β be any element of $A \setminus \mathbb{R}$. Then just as before we can find an element $\hat{i} \in A$ such that $\hat{i}^2 = -1$ and β is in the subalgebra generated by 1 and $\hat{i}$. However, then, since A is by assumption commutative, we have $(i - \hat{i})(i + \hat{i}) = i^2 - \hat{i}^2 = 0$, whence $\hat{i} = \pm i$. Hence $\hat{i}$, and therefore also β, belongs to A_0. Hence $A_0 = A$.

Exercise. Prove statement (b) of the theorem. □

The following corollary is simply a paraphrase of statement a) of the theorem.

Corollary. *The only fields containing $\mathbb{R}$ and of finite dimensions as vector spaces over $\mathbb{R}$ are $\mathbb{R}$ and $\mathbb{C}$.*

Exercise. Prove the following theorem.

Theorem 5. *There are no proper field extensions of $\mathbb{C}$ that are finite-dimensional as vector spaces over $\mathbb{C}$.*

(Note that this is equivalent to the "algebraic closure" of $\mathbb{C}$, i.e., to the fact that the irreducible polynomials over $\mathbb{C}$ are just those of degree 1: "the fundamental theorem of algebra.")

Exercise. Give an example of a field containing $\mathbb{R}$ (in other terminology, a commutative division algebra over $\mathbb{R}$) different from both $\mathbb{R}$ and $\mathbb{C}$.

Theorem 6 (Frobenius). *Any finite-dimensional division algebra A over $\mathbb{R}$ is isomorphic to $\mathbb{R}$, $\mathbb{C}$, or $\mathbb{H}$.*

Here is a sketch of the proof. If A is different from $\mathbb{R}$, then as in the proof of Theorem 4, we can find an element $i \in A$ such that $i^2 = -1$, and if, further, $A \neq \mathbb{C}$, then similarly one finds another element $\hat{\imath}$, say, such that $\hat{\imath}^2 = -1$ and $1, i, \hat{\imath}$ are linearly independent over $\mathbb{R}$. Using the argument beginning the proof of Theorem 4, we infer from the finite-dimensionality of A over $\mathbb{R}$ that every element of $A \setminus \mathbb{R}$ is a root of a quadratic equation with real coefficients. It can then be deduced—we omit the details—that $(i + \hat{\imath})^2 \in \mathbb{R}$, whence $i\hat{\imath} + \hat{\imath}i \in \mathbb{R}$; write $a := i\hat{\imath} + \hat{\imath}i$. Using the fact that a is real, we then have $(2\hat{\imath} + ai)^2 = 4\hat{\imath}^2 + 2a(i\hat{\imath} + \hat{\imath}i) + a^2 i^2 = -4 + a^2$, which must be negative, since otherwise $2\hat{\imath} + ai$ would be real, violating the linear independence of $1, i, \hat{\imath}$. Write $(2\hat{\imath} + ai)^2 = -b^2$, $b \in \mathbb{R}$. Setting $j := (2\hat{\imath} + ai)/b$, we have $j^2 = -1$ and $ij + ji = 0$. Finally, we set $k := ij$. We then have

$$ki = (ij)i = -(ji)i = -j(i^2) = j.$$

If k were linearly dependent on $1, i, j$, say $k = a + bi + cj$, right multiplication by i would yield $j = ai - b - ck$, whence $c^2 = -1$, giving a contradiction. Hence $1, i, j, k$ are linearly independent.

Consider the subalgebra

$$A_0 := \{a + bi + cj + dk \mid a, b, c, d \in \mathbb{R}\},$$

which, clearly, is isomorphic to $\mathbb{H}$. Suppose that $A_0 \neq A$. Then much as before we can find an element $l \in A \setminus A_0$, such that $il + li =: a \in \mathbb{R}$, $jl + lj =: b \in \mathbb{R}$, and $kl + lk =: c \in \mathbb{R}$, whence

$$lk = l(ij) = (li)j = (a - il)j = aj - i(lj)$$
$$= aj - i(b - jl) = aj - ib + kl,$$

yielding, on multiplying on the left by k, $(c - lk)k = ai - bj - l$, or $2l = ai - bj - ck \in A_0$. $\square$

8.3 Matrix algebras

Let $M_n(F)$ denote the set of all $n \times n$ matrices with entries from the field F. Under the usual addition and scalar multiplication of matrices, this set is a vector space of dimension $n \times n$, which the further operation of matrix multiplication turns into an associative algebra with multiplicative identity.

Exercise. Show that the set

$$\left\{ \begin{pmatrix} a & -b \\ b & a \end{pmatrix} \Big| \, a, b \in \mathbb{R} \right\}$$

is a subalgebra of $M_2(\mathbb{R})$ isomorphic to $\mathbb{C}$. Deduce the existence of an embedding $M_n(\mathbb{C}) \to M_{2n}(\mathbb{R})$. (Hence $M_{2n}(\mathbb{R})$ has a subalgebra isomorphic to $M_n(\mathbb{C})$.)

The importance of matrix algebras is made clear by the following result.

Theorem 7. *Every n-dimensional associative F-algebra A is isomorphic to a subalgebra of the matrix algebra $M_{n+1}(F)$. If A has a multiplicative identity element, then it is isomorphic to a subalgebra of $M_n(F)$.*

The first assertion of this theorem follows from the second via the following lemma, which we leave to the reader to verify.

Lemma 6. *Let A be any associative algebra over a field F, and consider the vector space*

$$\hat{A} := F \oplus A = \{ (\lambda, x) \mid \lambda \in F, \; x \in A \}.$$

Under the product operation defined by $(\lambda, x) \cdot (\mu, y) := (\lambda\mu, \lambda y + \mu x + xy)$, the space A becomes an associative algebra with identity. $\square$

The second statement of the theorem is proved as follows. With each element $a \in A$, we associate the map $T_a : A \to A$, defined by $x \longmapsto ax$ for all $x \in A$. It is easy to verify that the T_a are linear transfromations. Furthermore, if $a \neq b$, then since $T_a(1) = a \neq b = T_b(1)$, we have $T_a \neq T_b$. Write L_a for the matrix of the linear transformation T_a with respect to any fixed basis for A. By the preceding, the map $A \to M_n(F)$ sending each $a \in A$ to L_a is injective. It is an algebra homomorphism, since

$$L_{\lambda a + \mu b} = \lambda L_a + \mu L_b,$$
$$L_{ab}(x) = (ab)x = a(bx) = aL_b(x) = L_a L_b(x).$$

Hence we have the desired monomorphism from A to $M_n(F)$. $\square$

Exercise. Show that the maps from $\mathbb{H}$ to $M_2(\mathbb{C})$ and $M_4(\mathbb{R})$, defined respectively by

$$A_{\mathbb{C}} : u = a + bi + cj + dk \longmapsto \begin{pmatrix} a + bi & c + di \\ -c + di & a - bi \end{pmatrix},$$

$$A_{\mathbb{R}} : u = a + bi + cj + dk \longmapsto \begin{pmatrix} a & -b & c & -d \\ b & a & d & c \\ -c & -d & a & b \\ d & -c & -b & a \end{pmatrix},$$

define embeddings of the skew–field of quaternions as subalgebras of $M_2(\mathbb{C})$ and $M_4(\mathbb{R})$.

Note that either of these embeddings could be used to establish the associativity of multiplication of quaternions. They also have an interesting property relative to conjugation:

Exercise. Prove that the above maps satisfy $A(\bar{u}) = A(u)^*$, where in the real case $*$ denotes the transpose of the matrix, and in the complex case the conjugate–transpose.

Note that the matrices

$$\begin{pmatrix} 1 & 0 \\ 0 & 1 \end{pmatrix}, \begin{pmatrix} i & 0 \\ 0 & i \end{pmatrix}, \begin{pmatrix} 0 & 1 \\ -1 & 0 \end{pmatrix}, \begin{pmatrix} 0 & i \\ i & 0 \end{pmatrix},$$

which form a basis for the subalgebra of $M_2(\mathbb{C})$ isomorphic to $\mathbb{H}$ (via the embedding $A_\mathbb{C}$), are in physics called the "Pauli matrices."

8.4   Quaternions and rotations

In this section we shall restrict our attention to the "purely imaginary" quaternions, i.e., those of the form $u = bi + cj + dk$, naturally identifiable with the vectors of $\mathbb{R}^3$ expressed in terms of the standard basis i, j, k. The formula for multiplying two such quaternions, interpreted as vectors in this way, takes the form $uv = -u \cdot v + u \times v$, where now $u \cdot v$ denotes the Euclidean scalar, i.e., "dot," product, and $u \times v$ is the cross product of the vectors u and v. (Incidentally, one sees immediately from this that the purely imaginary quaternions do not form a subalgebra—although this is in any case completely obvious.)

Lemma 7 ([5]). *There is a natural one-to-one correspondence between the automorphisms of the $\mathbb{R}$-algebra of quaternions and the proper, origin–fixing isometries (i.e., rotations) of Euclidean $\mathbb{R}^3$.*

Let φ be an arbitrary automorphism of $\mathbb{H}$, considered as an $\mathbb{R}$-algebra, and let u, v, w be, in order, the images of i, j, k under φ. Since $\varphi(1) = 1$, we have $u^2 = \varphi(i)^2 = \varphi(i^2) = \varphi(-1) = -\varphi(1) = -1$. Write $u = a + u_1$, where a denotes the real part and u_1 the purely imaginary part of u. Then, since $u^2 = a^2 + 2au_1 - |u_1|^2 = -1$, we must have $au_1 = 0$, whence either $a = 0$ or $u_1 = 0$. Since $u_1 = 0$ would imply that $a^2 = -1$, which of course is not possible, it must be that $a = 0$. Hence u is purely imaginary, and, similarly, so are v and w. Since $w = \varphi(k) = \varphi(ij) = \varphi(i)\varphi(j) = uv$ and w is purely imaginary, we must have $u \cdot v = 0$ and $w = u \times v$, showing that the triple of vectors (u, v, w) constitutes a right-handed orthonormal basis for Euclidean $\mathbb{R}^3$. Thus the automorphism φ induces a proper (i.e., orientation-preserving) isometry of Euclidean $\mathbb{R}^3$.

Exercise. Show that a rotation of Euclidean $\mathbb{R}^3$ determines an automorphism of $\mathbb{H}$. □

Theorem 8. *The group $SO(3)$ consisting of the origin–fixing, proper isometries (which turn out to be just the rotations about axes through O) is isomorphic to* *the quotient group of the multiplicative group of unimodular quaternions (i.e., of norm 1) by the subgroup $\{\pm 1\}$.*

It is easy to check that for any nonzero quaternion α, the map defined by $x \longmapsto \alpha^{-1}x\alpha$ is an automorphism of $\mathbb{H}$, and is the same map as that corresponding to $\lambda\alpha$ for any nonzero real number λ. Hence as far as such automorphisms are concerned, we might as well assume $|\alpha| = 1$. Write $\alpha = a + u_0$, where a and u_0 are respectively the real and purely imaginary parts of α. Since $|\alpha|^2 = |a|^2 + |u_0|^2 = 1$, there is a unique angle θ, $0 \le \theta < 2\pi$, satisfying $a = \cos\theta$, $|u_0| = \sin\theta$, and then α may be written in the form

$$\alpha = a + u_0 = a + \frac{u_0}{|u_0|}|u_0| = \cos\theta + u\sin\theta,$$

where $|u| = 1$. (Here we are assuming $u_0 \ne 0$; if $u_0 = 0$, then $\sin\theta = 0$, and we may take u to be any purely imaginary quaternion of norm 1.) In either case it is easy to see that $\alpha u = u\alpha$, whence $\alpha^{-1}u\alpha = u$. Let v be any purely imaginary quaternion orthogonal to u, i.e., such that $u \cdot v = 0$, and set $w := uv$. We then have

$$\alpha^{-1}v\alpha = \bar{\alpha}v\alpha = (\cos\theta - u\sin\theta)v(\cos\theta + u\sin\theta)$$
$$= (v\cos\theta) - w\sin\theta)(\cos\theta + u\sin\theta)$$
$$= v(\cos^2\theta - \sin^2\theta) - 2w\sin\theta\cos\theta = v\cos 2\theta - w\sin 2\theta.$$

A similar calculation yields $\alpha^{-1}w\alpha = w\cos 2\theta + v\sin 2\theta$. Thus we see that the automorphism $\varphi_\alpha : x \longmapsto \alpha^{-1}x\alpha$ fixes u and rotates the plane spanned by the vectors v and w through the angle 2θ. Since every orientation-preserving orthogonal transformation of Euclidean $\mathbb{R}^3$ is a rotation about some axis through O, it follows that the map defined by $\alpha \longmapsto \varphi_\alpha$ is surjective, i.e., a homomorphism from the group of unimodular quaternions onto the group $SO(3)$.

Exercise. Verify that the set of unimodular quaternions forms a group under multiplication and that the map $\alpha \longmapsto \varphi_\alpha$ is indeed a homomorphism from this group onto the group $SO(3)$ of proper, origin-fixing motions of Euclidean $\mathbb{R}^3$. $\square$

Let us now calculate the kernel of this homomorphism. A unimodular quaternion α is in the kernel precisely if $\alpha^{-1}x\alpha = x$, i.e., $x\alpha = \alpha x$, for all $x \in \mathbb{H}$. Writing $\alpha = a + bi + cj + dk$, it follows by taking $x = i$ that $c = d = 0$, and by taking $x = j$ that $b = 0$. Hence $\alpha = a$, and since $|a| = |\alpha| = 1$, we infer that $\alpha = \pm 1$. $\square$

Exercise. Prove that every automorphism of the algebra of quaternions is inner, i.e., has the form φ_α.

Exercise. Prove that the group of unimodular quaternions is isomorphic to the group

$$SU(2) := \{A \in M_2(\mathbb{C}) \mid AA^* = I, \det A = 1\}.$$

We conclude the section by stating without proof two facts revealing a close connection between quaternions and the motions of 4-dimensional Euclidean space.

For this purpose we shall identify $\mathbb{H}$ with Euclidean $\mathbb{R}^4$, so that the Euclidean scalar ("dot") product is given by $u \cdot v = \mathrm{Re}(\bar{u}v) = \mathrm{Re}(u\bar{v})$.

Lemma 8. *If $\alpha \in \mathbb{H}$ has norm 1, then the operation of left (or right) multiplication of $\mathbb{H}$ by α yields an orthogonal transformation of $\mathbb{R}^4$.*

Theorem 9. *With each pair (β, γ) of unimodular quaternions, associate the (linear) transformation of $\mathbb{R}^4$ given by $u \longmapsto \beta^{-1} u \gamma$, $u \in \mathbb{H}$. This correspondence induces an isomorphism between the group $SO(4)$ of proper, origin-fixing motions of Euclidean $\mathbb{R}^4$ and the quotient of the direct product of the group of unimodular quaternions with itself, by its subgroup $\{\pm(1, 1)\}$.*

CHAPTER 9

The Derivative

9.1 Geometry and mechanics

Let $\mathbf{f}(t)$ be a function of a real variable t, $a < t < b$, taking as its values 3–dimensional real vectors. (Such a function is best visualized as giving the position–vector in 3–space of some particle at the various times t, so that the tip of the vector $\mathbf{f}(t)$ (with tail held fixed) describes the trajectory of the particle as the "time"–parameter t varies over (a, b).) In terms of a coordinate system in 3-space, we have $\mathbf{f}(t) = (f_1(t), f_2(t), f_3(t))$ at each t; the f_i are the *coordinate functions* of $\mathbf{f}$ relative to this coordinate system. On applying the usual definition of the derivative[1] of a real-valued function to the situation of a vector-valued function $\mathbf{f}(t)$, one obtains in terms of the coordinate functions simply $\mathbf{f}'(t) = (f_1'(t), f_2'(t), f_3'(t))$. Alternatively, one may simply take this as the definition, in which case it becomes incumbent upon one to do the following exercise:

Exercise. Show that coordinate–wise differentiation of a vector–function $\mathbf{f}(t)$ is independent of the choice of coordinate system.

It is not difficult to verify, either directly or via componentwise differentiation, that differentiation of vector–functions obeys the following familiar rules of differentiation.

Exercise. Show the following:

---

[1] *Translator's note.* The intended definition here is: $\mathbf{f}'(t) := \lim_{h \to 0} \frac{\mathbf{f}(t+h)-\mathbf{f}(t)}{h}$. This definition is, of course, motivated by the natural geometrical and kinematical interpretations of $\mathbf{f}'(t)$: geometrically, as the tangent vector to the parametrized curve defined by $\mathbf{f}(t)$ at the tip of this vector; and kinematically as the velocity vector of the aforementioned particle at time t.

(a) $(\mathbf{f} + \mathbf{g})' = \mathbf{f}' + \mathbf{g}'$.

(b) $(\varphi\mathbf{f})' = \varphi'\mathbf{f} + \varphi\mathbf{f}'$ (where φ is a scalar function).

(c) $(\mathbf{f} \cdot \mathbf{g})' = \mathbf{f}' \cdot \mathbf{g} + \mathbf{f} \cdot \mathbf{g}'$ (where $\cdot$ denotes the Euclidean scalar ("dot") product of vectors).

(d) $\mathbf{f} = \overrightarrow{\text{const.}}$ if and only if $\mathbf{f}'(t) = \mathbf{0}$ for all $t \in (a, b)$.

Note that the analogue for scalar functions of the last statement is proved using the mean–value theorem, which has no obvious counterpart in the case of vector functions. For example, taking $\mathbf{f}(t) := \cos t \, \mathbf{i} + \sin t \, \mathbf{j}$, we have $\mathbf{f}(0) = \mathbf{f}(2\pi)$, yet $\mathbf{f}'(t) = -\sin t \, \mathbf{i} + \cos t \, \mathbf{j} \neq \mathbf{0}$ for all $t \in (0, 2\pi)$. Thus to prove (d) one should use the componentwise formula for $\mathbf{f}'$.

The next formula will serve as a basic technical tool in the sequel.

Lemma 1. *One has*

$$\frac{d}{dt}|\mathbf{f}(t)| = \frac{\mathbf{f}(t) \cdot \mathbf{f}'(t)}{|\mathbf{f}(t)|} = \mathbf{f}'(t) \cdot \mathbf{e}(t),$$

where $\mathbf{e}(t)$ is the unit vector in the direction of $\mathbf{f}(t)$.

Here is the calculation in terms of coordinate functions. Choosing the coordinate system to be rectangular Cartesian, we have $|\mathbf{f}(t)| = \sqrt{f_1(t)^2 + f_2(t)^2 + f_3(t)^2}$. On differentiating this somewhat unwieldy function, we obtain

$$\frac{d}{dt}|\mathbf{f}(t)| = \frac{d}{dt}\left(\sqrt{f_1(t)^2 + f_2(t)^2 + f_3(t)^2}\right)$$

$$= \frac{2f_1 f_1' + 2f_2 f_2' + 2f_3 f_3'}{2\sqrt{f_1^2 + f_2^2 + f_3^2}} = \frac{\mathbf{f}(t) \cdot \mathbf{f}'(t)}{|\mathbf{f}(t)|}.$$

(The coordinate-free computation is rather simpler: one observes that $|\mathbf{f}(t)|^2 = \mathbf{f}(t)^2$ (i.e., $\mathbf{f}(t) \cdot \mathbf{f}(t)$), and differentiates using rule (c) above in conjunction with the ordinary "chain rule.") $\square$

The approach via coordinate functions is always available for us to use; however, computation tends to be simpler in terms of vectors, and moreover, the vector notation lends itself more to visualization or to making sense of the process at hand.

One of the basic theorems of the differential calculus is the following one (attributed to Fermat): If a function $\varphi(t)$ defined on an interval $[a, b]$ attains a greatest (or least) value at $t_0 \in (a, b)$ and is differentiable at that point, then $\varphi'(t_0) = 0$.

As an example of the use of this result, we shall prove the following consequence of it.

Corollary (of Fermat's theorem). *Let $\mathcal{E}$ be a subset of the plane, with boundary a simple, closed, differentiable curve C. Then any diameter AB of $\mathcal{E}$ (i.e., straight–line segment of greatest length having its endpoints A, B in $\mathcal{E}$) is perpendicular to the tangent lines to C at A and B.*

This assertion is intuitively perfectly obvious—as obvious, indeed, as Fermat's theorem itself.

However, here is a formal proof. Taking A to be the origin, let $\mathbf{f}(t)$, $t \in (a, b)$, be any differentiable parametrization of the curve $C \setminus \{A\}$; thus $\mathbf{f}(t)$ is a vector from A to a variable point M_t of C. Set $\varphi(t) := |AM_t| (= |\mathbf{f}(t)|)$, and suppose that $B = M_{t_0}$. Then since $\varphi(t_0)$ is the greatest value assumed by $\varphi(t)$, by Lemma 1 and Fermat's theorem we must have

$$\varphi'(t_0) = \mathbf{f}'(t_0) \cdot \frac{\overrightarrow{AB}}{|AB|} = 0,$$

so that the segment AB is perpendicular to the tangent vector $\mathbf{f}'(t_0)$ to C at the point B. $\square$

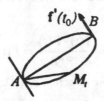

As noted above, the concept of a vector–function arises in mechanics as a means of specifying the position (relative to the origin) at various times t of, for instance, a point–particle, whose velocity $\mathbf{v}$ (i.e., instantaneous rate of change of position with respect to time) is then naturally given by the derivative, i.e., $\mathbf{v} := \mathbf{f}'$, and acceleration by $\mathbf{a} := \mathbf{v}' = \mathbf{f}''$.

Problem 1 (concerning a thrown stick).   Show that however a straight, rigid rod moves in space, the projections of the velocities of its ends onto the line in which it lies are at every instant equal.

This theorem of mechanics is, as we shall now see, a simple consequence of the basic formula of Lemma 1.

Denote by A_t and B_t the positions of the ends of the rod at time t, and set $\mathbf{f}(t) := \overrightarrow{OA_t}$, $\mathbf{g}(t) := \overrightarrow{OB_t}$, the position–vectors of the ends at time t, relative to some origin O; let $\mathbf{e}(t)$ be the unit vector in the direction of $\overrightarrow{AB}$ (see the diagram below). We wish to show that $\mathbf{e} \cdot \mathbf{v}_A = \mathbf{e} \cdot \mathbf{v}_B$ or $\mathbf{e} \cdot (\mathbf{v}_A - \mathbf{v}_B) = 0$, i.e., $\mathbf{e} \cdot (\mathbf{f}' - \mathbf{g}') = 0$, which is in fact true, since by Lemma 1, the left-hand side is equal to $\frac{d}{dt} |\mathbf{f} - \mathbf{g}|$, and this is identically zero by virtue of the fact that $|\mathbf{f} - \mathbf{g}|$ is just the length of the rod, and so constant.

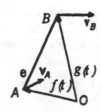

Problem 2. Prove Snell's law governing the bending of a light–ray in passing from one homogeneous medium to another: The ratio of the sines of the angle (with the vertical) at which the ray meets the boundary between the two media and the angle at which it leaves the boundary is equal to the ratio of the speeds of light in the respective media.

Clearly, before we can convert this problem into mathematical form, we need to know something appropriate about the behavior of light. The physical law applying here is the very general "principle of least action," which in the present context specializes to "Fermat's principle": A ray of light passing from a point A to a point B always takes a path minimizing the time of transit.

Clearly, since the path of the light–ray in each medium is straight, the whole path from A to B lies in a plane. Denote by M_t a variable point on the curve of intersection of this plane (i.e., the plane in which the light ray travels) with the boundary between the two media (see the diagram below); we assume that the point M_t moves along this curve as the time t varies. For each t denote by $\varphi(t)$ the time it would take for a light ray to travel along the path AM_tB, made up of the two straight–line segments AM_t and M_tB. Define $\mathbf{f}(t) := \overrightarrow{AM_t}$ and $\mathbf{g}(t) := \overrightarrow{M_tB}$. Denoting by c_1 and c_2 the speeds of light in the respective media, we have

$$\varphi(t) = \frac{|\mathbf{f}(t)|}{c_1} + \frac{|\mathbf{g}(t)|}{c_2}.$$

Since $\mathbf{f}(t) + \mathbf{g}(t) \equiv \overrightarrow{AB} = \text{const.}$, we have $\mathbf{f}' = -\mathbf{g}' = \mathbf{v}$, say, the velocity of the point M_t relative to A. Hence writing $\mathbf{e}_1$ and $\mathbf{e}_2$ for the unit vectors in the directions of AM_t and M_tB respectively (as in the diagram), we have

$$\varphi' = \frac{\mathbf{e}_1 \cdot \mathbf{f}'}{c_1} + \frac{\mathbf{e}_2 \cdot \mathbf{g}'}{c_2} = \mathbf{v} \cdot \left(\frac{\mathbf{e}_1}{c_1} - \frac{\mathbf{e}_2}{c_2} \right).$$

According to Fermat's principle, the actual path taken will be such as to minimize φ, and hence by Fermat's theorem, this will occur when $\varphi' = 0$, i.e., when the angle of incidence θ_1 and angle of refraction θ_2 satisfy

$$\frac{\sin \theta_1}{\sin \theta_2} = \frac{\mathbf{v} \cdot \mathbf{e}_1}{\mathbf{v} \cdot \mathbf{e}_2} = \frac{c_1}{c_2}.$$

Problem 3. Establish the following optical property of the ellipse: Any tangent to an ellipse makes equal angles with the line–segments joining the point of contact to the foci of the ellipse.

(Use the defining property $|F_1M| + |F_2M| = $ const. for an ellipse, together with $\overrightarrow{F_1M} + \overrightarrow{F_2M} = \overrightarrow{F_1F_2} = $ const.—see the diagram.)

Thus we see (as earlier in the corollary to Fermat's theorem) that the above scheme for applying the derivative is often effective also in the solution of geometrical problems; here is a further example.

Problem 4. Given an acute–angled triangle, let P_0 be a point such that the sum of the distances from P_0 to the vertices of the triangle is least. Prove that each side of the triangle subtends an angle of 120° at P_0.

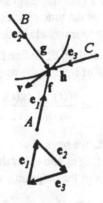

Consider a variable point P_t moving in time along a curve passing through such a point $P_0 = P_{t_0}$. Suppose to begin with that P_0 does not coincide with any of the vertices A, B, C, of the given triangle. Denoting by $\mathbf{f}, \mathbf{g}, \mathbf{h}$ the vectors directed from A, B, C respectively, to P_0 (as in the diagram), we see that the velocity of P_0 satisfies $\mathbf{v} = \mathbf{f}'(t) = \mathbf{g}'(t) = \mathbf{h}'(t)$. Write

$$\varphi(t) := |\mathbf{f}(t)| + |\mathbf{g}(t)| + |\mathbf{h}(t)|.$$

By Fermat's theorem, we have $\varphi'(t_0) = 0$, whence $\mathbf{v} \cdot (\mathbf{e}_1 + \mathbf{e}_2 + \mathbf{e}_3) = 0$, where $\mathbf{e}_1, \mathbf{e}_2, \mathbf{e}_3$ are the appropriate unit vectors. Hence $\mathbf{e}_1 + \mathbf{e}_2 + \mathbf{e}_3 = \mathbf{0}$ (why?). Since $\mathbf{e}_1, \mathbf{e}_2, \mathbf{e}_3$ have the same length, it follows that the angle between each pair of them is 120°.

Exercise. Where in the above argument was the assumption $P_0 \neq A, B, C$ used?

Now, if P is a point in the interior of the side AB of our (acute–angled) triangle, close to A, then it is easy to see that $PA + PB + PC < AB + AC$; hence $P_0 \neq A$ and, similarly, $P_0 \neq B, C$.

This solution shows both the strength and the weakness of the technique we have been using. The weakness consists in the need to assume at the outset that there *is* a minimum point, i.e., a point where the function in question attains a minimum value, and moreover, that the function is differentiable at that point.

We shall remedy the first of these defects in the solution of the preceding problem, by proving that the sum $\psi(P) := PA + PB + PC$ has a least value. Clearly such a point, if it exists, could not lie outside some sufficiently large disk containing the triangle, for instance one of radius equal to the perimeter of the triangle. Since the function ψ is continuous, and the disk is closed and bounded, and therefore compact, we have by Weierstrass's theorem (in generalized form) that it does attain a least value at some point of the disk.

Exercise. Try to generalize the assertion of Problem 4 to the situation of an arbitrary triangle.

9.2   Functional equations

Problem 5. Find all functions f satisfying:
  (a) $f(x + y) = f(x) + f(y)$ for all $x, y \in \mathbb{R}$.
  (b) $f(x + y) = \frac{f(x)+f(y)}{1-f(x)f(y)}$ for all admissible $x, y \in \mathbb{R}$.

Clearly, any function $f(x) := cx$ satisfies (a); it turns out that there are no other continuous such functions (see the second exercise below). Here we shall solve both parts of the problem under the additional assumption that the function f has a derivative at zero.

Observe firstly that on setting $x = y = 0$ in (a) and (b), one obtains in each case $f(0) = 0$. For a function satisfying (a) (and differentiability at 0), we have

$$\lim_{y \to 0} \frac{f(x+y)-f(x)}{y} = \lim_{y \to 0} \frac{f(y)}{y} = \lim_{y \to 0} \frac{f(y)-f(0)}{y} = f'(0),$$

so that f is differentiable everywhere, moreover with constant derivative $f'(x) \equiv c$, say. Hence $f(x) = cx + d$, and since $f(0) = 0$, we have $d = 0$, so $f(x) = cx$.

For a function f satisfying condition (b) (and differentiability at 0, say $f'(0) = c$), we have

$$f'(x) = \lim_{y \to 0} \frac{f(x+y)-f(x)}{y} = \lim_{y \to 0} \left(\frac{1}{y} \left(\frac{f(x)+f(y)}{1-f(x)f(y)} - f(x) \right) \right)$$

$$= \lim_{y \to 0} \left(\frac{f(y)}{y} \frac{1+f^2(x)}{1-f(x)f(y)} \right) = f'(0)(1+f^2(x)) = c(1+f^2(x)).$$

It follows that

$$(\arctan f(x))' = \frac{f'(x)}{1+f^2(x)} = c,$$

whence $\arctan f(x) = cx + d$, and then since $f(0) = 0$, we have $d = 0$. Thus $f(x) = \tan cx$. As always, it remains to verify that the function one has found does actually satisfy the original functional equation; this is necessary since the differential equation obtained from the functional equation is, on the face of it, only a consequence of the latter.

Exercise. Find all differentiable functions satisfying $f(x + y) = f(x) + f(y) + 2xy$ for all $x, y \in \mathbb{R}$.

Exercise. Prove that if an arbitrary function f satisfies $f(x + y) = f(x) + f(y)$ for all real x, y, then (a) $f(n) = nf(1)$ for all integers n; (b) $f(m/n) = (m/n)f(1)$ for all rational numbers m/n; (c) if f is continuous at zero, then $f(x) = xf(1)$ for all real x.

Problem 6. Show that a function f satisfying $|f(x) - f(y)| \le (x - y)^2$ for all $x, y \in \mathbb{R}$ must be constant.

Since $\left| \dfrac{f(y) - f(x)}{y - x} \right| \le |x - y| \to 0$ as $y \to x$, we have $f'(x) = 0$ for all x. Here is an another solution not using the derivative: Since

$$|f(x) - f(0)| = \left| \sum_{k=1}^{n} (f(\tfrac{kx}{n}) - f(\tfrac{(k-1)x}{n})) \right|$$

$$\le \sum_{k=1}^{n} |f(\tfrac{kx}{n}) - f(\tfrac{(k-1)x}{n})| \le \sum_{k=1}^{n} \tfrac{x^2}{n^2} = \tfrac{x^2}{n} \to 0 \text{ as } n \to 0,$$

it follows that $f(x) = 0$ for all $x \in \mathbb{R}$.

9.3   The motion of a point–particle

In this section we turn again to problems in mechanics. As noted earlier, we shall in the standard way consider the trajectory (or path) of a point–particle to be the curve traced out as the time t varies by the tip of $\mathbf{f}(t)$ with its tail held fixed at the origin (here $\mathbf{f}$ is an appropriate vector–function). Thus the trajectory is the family of points $\{M_t \mid \overrightarrow{OM_t} = \mathbf{f}(t), t \in (a, b)\}$.

Problem 7. The motion of an electron in a constant magnetic field is described by a vector function $\mathbf{f}$ satisfying the equation $\mathbf{f}'' = \mathbf{f}' \times \mathbf{H}$, where $\mathbf{H}$ is the magnetic field strength (assumed constant) and $\times$ denotes the cross product. Show that the electron's trajectory is a circular helix.

The bulk of the solution is contained in the following lemma.

Lemma 2. *A vector–function* $\mathbf{v}$ *satisfying* $\mathbf{v}' = \mathbf{v} \times \mathbf{c}$, *where* $\mathbf{c}$ *is a constant vector, describes the trajectory of a point–particle moving with constant speed on a circle contained in a plane perpendicular to* $\mathbf{c}$, *i.e.,* $\mathbf{v}(t) = \mathbf{v}_0 + a(\mathbf{i} \cos(\alpha t + \beta) + \mathbf{j} \sin(\alpha t + \beta))$, *where the vector* $\mathbf{v}_0$ *is parallel to* $\mathbf{c}$, *and the unit vectors* $\mathbf{i}, \mathbf{j}$ *are perpendicular to* $\mathbf{c}$.

From $\mathbf{v}' = \mathbf{v} \times \mathbf{c}$ it follows firstly that $\mathbf{v}' \cdot \mathbf{c} = 0$, whence $\mathbf{v} \cdot \mathbf{c} = $ const., and secondly that $\mathbf{v}' \cdot \mathbf{v} = 0$, whence $|\mathbf{v}| = $ const. Hence the angle between the vectors $\mathbf{v}$ and $\mathbf{c}$ is also constant. It follows that $\mathbf{v}$ has the form

$$\mathbf{v}(t) = \mathbf{v}_0 + a(\mathbf{i} \cos\theta(t) + \mathbf{j} \sin\theta(t)),$$

where $\mathbf{v}_0$ is a constant vector parallel to $\mathbf{c}$. (Verify this, first writing $\mathbf{v}(t)$ in the form $v_1(t)\mathbf{c} + v_2(t)\mathbf{i} + v_3(t)\mathbf{j}$.) Hence

$$|\mathbf{v}'(t)| = |-a(\mathbf{i}\sin\theta(t) - \mathbf{j}\cos\theta(t))\theta'(t)| = |a||\theta'(t)|.$$

Now, using $\mathbf{v}' = \mathbf{v} \times \mathbf{c}$ again, together with the constancy of $|\mathbf{v}|$ and the angle between $\mathbf{v}$ and $\mathbf{c}$, we infer that $|\mathbf{v}'| = $ const. Hence $\theta'(t) = $ const., whence $\theta(t) = \alpha t + \beta$. $\square$

Turning now to the problem, we have that the velocity $\mathbf{v}$ of the electron satisfies $\mathbf{v}' = \mathbf{v} \times \mathbf{H}$, so that it can be expressed as in the above lemma, with $\mathbf{H}$ in the role of $\mathbf{c}$. On integrating (i.e., antidifferentiating) that expression, we obtain

$$\mathbf{f}(t) = \mathbf{f}_0 + \mathbf{v}_0 t + \frac{a}{\alpha}(\mathbf{i}\sin(\alpha t + \beta) - \mathbf{j}\cos(\alpha t + \beta)),$$

which is a vector parametrization of a circular helix.

In what follows we shall need to use the following further rule of differentiation.

Exercise. Prove that $(\mathbf{f} \times \mathbf{g})' = \mathbf{f}' \times \mathbf{g} + \mathbf{f} \times \mathbf{g}'$.

Problem 8. The trajectory of a point–mass moving in the gravitational field of a massive body situated at the origin is given by a vector–function $\mathbf{f}$ satisfying $\mathbf{f}'' = -\mu\mathbf{f}/|\mathbf{f}|^3$ for some constant μ (Newton's universal law of gravitation). Deduce that (a) such a trajectory must lie in some plane; (b) any curve homothetic to some such trajectory is itself a possible trajectory; (c) given two mutually homothetic trajectories, the ratio of the squares of their respective periods is equal to the cube of the factor of dilation.

(The assertions (b) and (c) represent two of the three famous laws of Kepler governing the motion of the planets around the sun; the missing law states that the trajectories are ellipses—more generally, conic sections.)

We give only hints as to the solutions. For (a), show that the vector $\mathbf{f} \times \mathbf{f}'$ is constant. For (b) and (c), prove that if $\mathbf{f}(t)$ satisfies the equation, then so does $\mathbf{g}(t) := k\mathbf{f}(k^{-3/2}t)$.

As is taught in highschool courses in physics, if a point moves with constant speed v on a circle of radius R, then its acceleration is at all times directed towards the center of the circle, and the magnitude of this "centripetal" acceleration is given by $w_N = v^2/R$. We shall now establish a much more general formula pertaining to the completely general situation of a point moving along an arbitrary space curve.

Theorem 1 (on the components of acceleration). *Let $\mathbf{f}(t)$ be a vector function describing the motion of a point along a curve C, and write $\mathbf{a} := \mathbf{f}''$, the acceleration. At each time t one may write $\mathbf{a}(t) = \mathbf{a}_T(t) + \mathbf{a}_N(t)$, where $\mathbf{a}_T$ is the component* *of* **a** *tangential to the curve at the tip of* $\mathbf{f}(t)$ *(i.e., in the direction of the velocity* $\mathbf{v}(t)$*), and* $\mathbf{a}_N$ *the component normal to the curve. Then at each time t, one has*

$$\mathbf{a}_T = \frac{d|\mathbf{v}|}{dt} \cdot \frac{\mathbf{v}}{|\mathbf{v}|}, \quad |\mathbf{a}_N| = \frac{|\mathbf{v}|^2}{R},$$

where R is the "radius of curvature" of the curve C at the tip of $\mathbf{f}(t)$ *(and so depends only on the curve, i.e., is independent of the speed at which the point traverses it).*

Writing $\mathbf{a}_T = \lambda \mathbf{v}/|\mathbf{v}|$, we have

$$\lambda = \frac{\mathbf{a}_T \cdot \mathbf{v}}{|\mathbf{v}|} = \frac{\mathbf{a} \cdot \mathbf{v}}{|\mathbf{v}|} = \frac{\mathbf{v}' \cdot \mathbf{v}}{|\mathbf{v}|} = \frac{d|\mathbf{v}|}{dt},$$

where the last equality follows from Lemma 1. Turning to $\mathbf{a}_N$, we have

$$|\mathbf{a}_N|^2 = |\mathbf{a}|^2 - |\mathbf{a}_T|^2 = |\mathbf{v}'|^2 - \frac{|\mathbf{v}' \cdot \mathbf{v}|^2}{|\mathbf{v}|^2}$$

$$= \frac{|\mathbf{v}|^2 |\mathbf{v}'| \left(1 - \cos^2 (\widehat{\mathbf{v}, \mathbf{v}'})\right)}{|\mathbf{v}|^2} = \frac{|\mathbf{v}|^2 |\mathbf{v}'| \sin^2 (\widehat{\mathbf{v}, \mathbf{v}'})}{|\mathbf{v}|^2},$$

whence

$$|\mathbf{a}_N| = \frac{|\mathbf{v} \times \mathbf{v}'|}{|\mathbf{v}|} = |\mathbf{v}|^2 \frac{|\mathbf{v} \times \mathbf{v}'|}{|\mathbf{v}|^3}.$$

It remains to show that the quantity

$$\frac{|\mathbf{v} \times \mathbf{v}'|}{|\mathbf{v}|^3} = \frac{|\mathbf{f}' \times \mathbf{f}''|}{|\mathbf{f}'|^3},$$

which it turns out to be appropriate to call the "curvature" of the curve C, is independent of the parametrization of C.

Lemma 3. *If* **f** *and* **g** *are vector–functions, and* φ *a scalar function with nowhere-vanishing derivative such that* $\mathbf{f}(t) = \mathbf{g}(\varphi(t))$, *then*

$$\frac{|\mathbf{f}' \times \mathbf{f}''|}{|\mathbf{f}'|^3} = \frac{|\mathbf{g}' \times \mathbf{g}''|}{|\mathbf{g}'|^3}.$$

(Differentiate the identity $\mathbf{f}(t) = \mathbf{g}(\varphi(t))$ twice.) $\square$ $\square$

The acceleration $\mathbf{a} := \mathbf{f}''$, which figured prominently in this section, is of especial interest, of course, because of Newton's second law of motion $\mathbf{F} = m\mathbf{a}$, equating the net force acting on a mass–point with its mass times its acceleration.

In the remainder of this chapter we shall for the most part be considering "systems with one degree of freedom," or, to put it more mathematically, differential equations involving only scalar functions.

9.4   On the number e

It is clear from the above how to solve the simplest differential equations, namely $y' = k$ and $y'' = k$, $k =$ const. We now proceed to more difficult cases.

Problem 9. Prove that the solutions of the equation $y' = ky$ are all of the form $y = f(x) = Ce^{kx}$.

One first verifies—by differentiating it—that such a function is a solution. To prove the converse, we let $f(x)$ be any solution, and differentiate the product $f(x)e^{-kx}$:

$$\frac{d}{dx}(f(x)e^{-kx}) = f'(x)e^{-kx} - kf(x)e^{-kx} = (f'(x) - kf(x))e^{-kx} = 0.$$

Hence $f(x)e^{-kx} = C =$ const., whence $f(x) = Ce^{kx}$.

The following lemma concerns equations of order 2, i.e., involving the second derivative.

Lemma 4. Let $f(x)$ and $g(x)$ be arbitrary solutions of the equations $y'' - k^2 y = 0$ and $y'' + \omega^2 y = 0$ respectively. Then $f(x) = ae^{kx} + be^{-kx}$ and $g(x) = a\cos\omega x + b\sin\omega x$.

For the first of these equations, consider the following differentiation:

$$((f'(x) \pm kf(x))e^{\mp kx})' = (f''(x) \pm kf'(x))e^{\mp kx} \mp (f'(x) \pm kf(x))ke^{\mp kx}$$
$$= e^{\mp kx}(f''(x) - k^2 f(x)) = 0.$$

Hence

$$\begin{cases} f'(x) + kf(x) = c_1 e^{kx}, \\ f'(x) - kf(x) = c_2 e^{-kx}, \end{cases}$$

whence the formula for $f(x)$ follows easily.

Exercise. Establish the general form of the solution of the second equation in the lemma. $\square$

The reader will have noticed that in the solutions of some of these elementary differential equations the fundamental number of mathematical analysis appears, namely e, usually defined as the limit of a certain sequence ($e = \lim_{n\to\infty}(1+1/n)^n$) or the sum of a certain series ($e = \sum_{k=0}^{\infty} 1/k!$). Consider the solution $f(x)$ of the differential equation $y' = y$ satisfying the initial condition $f(0) = 1$; by Problem 9 one has $f(1) = e$. It turns out that the terms $(1 + 1/n)^n$ of the above sequence arise via the successive approximations to the solution of this differential equation by means of the polygonal arcs occuring in Euler's method, while the partial sums $\sum_{k=0}^{n} 1/k!$ arise from the Picard method of successive approximations.

Exercise. (a) Consider the polygonal arc $P_0 P_1 \cdots P_n$ in the plane, with coordinates of the vertices P_i respectively $(0, 1), (1/n, y_1), (2/n, y_2), \ldots (1, y_n)$, where y_i is defined recursively by $y_{i+1} = y_i + y_i(x_{i+1} - x_i)$. (Note that this arises from the mean–value theorem in the form $y_{i+1} = y_i + y'(c)(x_{i+1} - x_i)$, $x_i < c < x_{i+1}$, with $y'(c) = y(c)$ approximated by y_i). Show that $y_n = (1 + 1/n)^n$.

(b) Consider the sequence of functions $\varphi_i(x)$ defined recursively by $\varphi_0(x) \equiv 1$ and

$$\varphi_{i+1}(x) = 1 + \int_0^x \varphi_i(u)du, \quad i = 0, 1, \ldots.$$

Prove that $\varphi_n(1) = \sum_{k=0}^n 1/k!$.

(In (b) one has, in fact, $\varphi_n(x) = \sum_{k=0}^n x^k/k!$.)

It is often more convenient to consider an appropriate *integral* equation in place of a given differential equation.

Exercise. Prove that if φ is a continuous function satisfying the integral equation $\varphi(x) = 1 + \int_0^x \varphi(u)du$, then φ is also a solution of the differential equation $y' = y$ (in fact, *the* solution satisfying $\varphi(0) = 1$).

Theorem 2. *The function* $\varphi(x) := \sum_{k=0}^{\infty} x^k/k!$ *is a solution of the differential equation* $y' = y$, *in fact, the unique solution satisfying* $\varphi(1) = e := \sum_{k=0}^{\infty} 1/k!$.

To see this, write $\varphi(x) = \sum_{k=0}^{i} x^k/k! + \sum_{k=i+1}^{\infty} x^k/k! = \varphi_i(x) + r_i(x)$, say. Then

$$\left| \int_0^x \varphi(u)du - \int_0^x \varphi_i(u)du \right| = \left| \int_0^x \left(\sum_{k=i+1}^{\infty} \frac{u^k}{k!} \right) du \right|$$

$$\leq \left| \int_0^x \left(\sum_{k=i+1}^{\infty} \frac{|x|^k}{k!} \right) du \right|$$

$$= |x| \sum_{k=i+1}^{\infty} \frac{|x|^k}{k!} \to 0 \text{ as } i \to \infty,$$

where in the last step we have used the fact that the power series defining φ converges (for all x). Hence for each x we have $\int_0^x \varphi_i \to \int_0^x \varphi$ as $i \to \infty$. Since by the penultimate exercise the φ_i satisfy the recurrence relation $\varphi_{i+1}(x) = 1 + \int_0^x \varphi_i(u)du$, it follows that $\varphi(x) = 1 + \int_0^x \varphi(u)du$. Hence by the preceding exercise φ satisfies the differential equation $y' = y$.

Exercise. Prove that the function φ is continuous everywhere. □

We now turn to Euler's method. The following lemma shows that in connection with this method also, one can write down a certain relevant integral equation. We first define a sequence of functions $\chi_n : [0, 1] \to [0, 1]$, by means of the formula $\chi_n(x) := \frac{1}{n}[nx]$, where [ ] denotes the "integer part" of a real number.

Lemma 5. *Let* ψ_n *denote the piecewise–linear function with graph the polygonal arc* $P_0 P_1 \cdots P_n$ *introduced in the above exercise on Euler's method. Then*

$$\psi_n(x) = 1 + \int_0^x \psi_n(\chi_n(u))du, \quad x \in [0, 1].$$

We use induction. Suppose inductively that the equation of the lemma holds for $0 \le x \le k/n$ for some nonnegative integer $k < n$, and consider $x \in [k/n, (k+1)/n)$. For any such x we have (see the above–mentioned exercise)

$$\psi_n(x) = y_k + y_k'(x - k/n) = \psi_n(k/n) + \psi_n'(k/n)(x - k/n)$$

$$= \psi_n(k/n) + \int_{k/n}^{x} \psi_n(\chi_n(u))du$$

$$= 1 + \int_0^{k/n} \psi_n(\chi_n(u))du + \int_{k/n}^{x} \psi_n(\chi_n(u))du$$

$$= 1 + \int_0^{x} \psi_n(\chi_n(u))du.$$

Since on both sides of the final equation the functions appearing are continuous, that equation must then hold also for $x = (k+1)/n$. $\square$

9.5 Contracting maps

Problem 10. Let $f : \mathbb{R} \to \mathbb{R}$ be a differentiable function satisfying $|f'(x)| \le q < 1$ for all $x \in \mathbb{R}$. Prove that the equation $f(x) = x$ has exactly one solution.

First the uniqueness: If $f(x) = x$ and $f(y) = y$, with $x \ne y$, then

$$|x - y| = |f(x) - f(y)| = |f'(c)(x - y)| < |x - y|,$$

a contradiction. The existence of a fixed point is established as follows: Suppose for definiteness that $f(0) > 0$, and let a be any number satisfying $a > f(0)/(1 - q)$. By the mean–value theorem, there is a number $b \in (0, a)$ such that $f(a) - f(0) = f'(b)a$, whence $f(a) - a \le f(0) + qa - a < 0$. Thus the function $g(x) := f(x) - x$, which is certainly continuous, is positive at $x = 0$ and negative at $x = a$. Hence by the intermediate–value theorem, $g(x_*) = 0$, i.e., $f(x_*) = x_*$, for some number $x_* \in (0, a)$.

(It is not dificult to see that in fact the fixed point x_* satisfies the estimate $|x_*| \le |f(0)|/(1 - q)$.)

Exercise. Give an example of a function showing that the condition $|f'(x)| < 1$ for all $x \in \mathbb{R}$ does not in general suffice for the existence of a fixed point.

The above argument established only that a fixed point must exist, without any indication as to how one might actually find it. However, there is an algorithm by means of which the solution of the equation $f(x) = x$ can be computed to any desired degree of accuracy. We shall formulate the result in question in a rather more general context, requiring the following preliminary definitions.

A *metric space* is by definition a set X endowed with a *metric* (to be thought of as giving the "distance" between pairs of points), i.e., a function $d : (x, y) \longmapsto d(x, y) \in \mathbb{R}$, $x, y \in X$, with the following three properties (the most elemental properties of "distance"): (1) $d(x, y) \ge 0$, and $d(x, y) = 0$ if and only if $x = y$;

(2) $d(x, y) = d(y, x)$; (3) $d(x, y) \leq d(x, z) + d(z, y)$. The last of these is called—for the obvious reason—the "triangle inequality." (The reader may, if he or she so wishes, simply take $X = \mathbb{R}$ with the usual distance $d(x, y) := |x - y|$, or, more generally, $X = \mathbb{R}^n$ with the usual Euclidean distance—these are in any case the prototypical examples.)

A metric space (X, d) is said to be *complete* if every "Cauchy sequence" of points from X has a limit (see Chapter 10). It is a basic result proved in courses in mathematical analysis that Euclidean $\mathbb{R}^n$ is complete.

Finally, a map $f : X \to X$ from a metric space to itself is called *contracting* if there exists a number $\alpha \in (0, 1)$ such that $d(f(x), f(y)) \leq \alpha d(x, y)$ for all $x, y \in X$.

Theorem 3 (Banach's fixed–point theorem for contracting maps).   *Every contracting map $f : X \to X$ of a complete metric space X has a unique fixed point, i.e., there is exactly one point $x_* \in X$ such that $f(x_*) = x_*$. Furthermore, starting from any point $x_0 \in X$ and defining recursively $x_{n+1} := f(x_n)$ for $n \geq 0$, one has $x_* = \lim_{n \to \infty} x_n$.*

The uniqueness is easy to see—along the lines of the argument used in the solution of Problem 10.

For the existence of a fixed point, consider any sequence $\{x_n\}_{n=0}^{\infty}$, where, as in the theorem, x_0 is an arbitrary point of X, and $x_{n+1} := f(x_n)$ for $n \geq 0$. Since for $n \geq 1$ we have

$$d(x_{n+1}, x_n) = d(f(x_n), f(x_{n-1})) \leq \alpha d(x_n, x_{n-1}),$$

where $\alpha \in (0, 1)$ is the contraction factor, it follows that $d(x_{n+1}, x_n) \leq \alpha^n d(x_1, x_0) = a\alpha^n$, where $a := d(x_1, x_0)$. Hence for $0 < m < n$ we have

$$d(x_m, x_n) \leq \sum_{i=m+1}^{n} d(x_{i-1}, x_i) \leq \sum_{i=m+1}^{n} a\alpha^{i-1} \leq \sum_{i=m}^{\infty} a\alpha^i = \frac{a\alpha^m}{1 - \alpha},$$

whence we infer that $d(x_m, x_n) \to 0$ as $m, n \to \infty$, i.e., the sequence $\{x_n\}_0^{\infty}$ is Cauchy, and so by the completeness has a limit x_*, say. We then have

$$f(x_*) = f(\lim_{n \to \infty} x_n) = \lim_{n \to \infty} f(x_n) = \lim_{n \to \infty} x_{n+1} = x_*,$$

so that x_* is indeed the fixed point. $\square$

Exercise. Prove that

$$d(x_*, x) \leq (1 - \alpha)^{-1} d(f(x), x) \quad \forall x \in X.$$

Corollary. *Let x_* be the fixed point of a contracting map f of X, with contraction factor α, and let g be any map of X to itself with a fixed point y_*. Then*

$$d(x_*, y_*) \leq (1 - \alpha)^{-1} \sup_{x \in X} d(f(x), g(x)).$$

By the preceding exercise,

$$d(x_*, y_*) \leq \frac{d(f(y_*), y_*)}{1 - \alpha} \leq \frac{d(f(y_*), g(y_*))}{1 - \alpha} \leq (1 - \alpha)^{-1} \sup_{x \in X} d(f(x), g(x)). \quad \square$$

An easy application of the mean–value theorem yields (see the solution of Problem 10) that a function $f : \mathbb{R} \to \mathbb{R}$ is contracting if $|f'(x)| \leq q < 1$ for all $x \in \mathbb{R}$. There is an analogous result for maps $f : \mathbb{R}^n \to \mathbb{R}^n$:

Exercise. Prove that a map $f : \mathbb{R}^n \to \mathbb{R}^n$ is contracting (with respect to the usual Euclidean metric) if there exists a constant q, $0 < q < 1$, such that for all $x \in \mathbb{R}^n$, the partial derivatives are bounded as follows:

$$\left| \frac{\partial f_i}{\partial x_j}(x) \right| \leq \frac{q}{n^2}, \ i, j = 1, 2, \ldots, n, \ 0 < q < 1.$$

Although the proof of the fixed-point theorem for contracting maps is, as we have seen, quite simple, its applications are both various and deep.

Problem 11. Show that for each positive real number a, the sequence defined by the recursion $x_{n+1} := \frac{1}{2}(x_n + a/x_n)$, with $x_0 \geq a$ but otherwise arbitrary, has as its limit $\sqrt{a}$.

(The function defined by $f(x) := \frac{1}{2}(x + a/x)$ restricts to a contracting map of the ray $[\sqrt{a}, +\infty)$.)

The next problem can be solved by a direct appeal to Theorem 3. Try to find an alternative elementary proof.

Problem 12. A geographical map of a certain region drawn to the scale of $1 : 10^5$ is placed on top of another map of the same region of scale $1 : 10^6$. Show that there is a point of the region such that the points corresponding to it on the maps lie exactly one over the other.

The fixed-point theorem for contracting maps can also be used to establish the existence and uniqueness of solutions of differential equations.

Denote by $C[a, b]$ the space of all functions continuous on the closed interval $[a, b]$, endowed with the metric given by

$$d_C(u, v) := \max_{t \in [a,b]} |u(t) - v(t)|.$$

This metric space is known to be complete. Consider now by way of example the differential equation $y' = y$, and the maps f, g_n defined on $C[0, a]$ by

$$f(u)(t) := 1 + \int_0^t u(\tau)d\tau, \quad g_n(u)(t) := 1 + \int_0^t u(\chi_n(\tau))d\tau.$$

Exercise. Prove that if $0 < a < 1$, then the maps $f, g_n : C[0, a] \to C[0, a]$ are contracting, and that

$$\max_{C[0,a]} |f(u) - g_n(u)| \to 0 \text{ as } n \to \infty.$$

Theorem 3 will have an important role to play also in what follows.

9.6 Linearization

For the sake of simplicity, we shall in this section confine ourselves to ordinary real-valued functions; however, the arguments given here may all without great difficulty be extended to the general situation of functions $f : \mathbb{R}^n \to \mathbb{R}^n$.

Instead of the usual highschool definition of the derivative, we shall use the following (equivalent) one, for many purposes more convenient (in particular, for extending the definition to the more general case of functions $f : \mathbb{R}^m \to \mathbb{R}^n$).

A number a is called the *derivative* of $f : \mathbb{R} \to \mathbb{R}$ at the point x_0 ($a = f'(x_0)$) if for all points $x_0 + h$ in some neighborhood of x_0, we can express $f(x_0 + h)$ in the form $f(x_0 + h) = f(x_0) + ah + \alpha(x_0, h)$, where $\frac{\alpha(x_0, h)}{h} \to 0$ as $h \to 0$. (For brevity, we shall henceforth write $\alpha(h)$ instead of $\alpha(x_0, h)$.)

For the most part the functions we consider will not just be differentiable, but will have continuous derivatives, i.e., will belong to the class C^1.

Lemma 6. *If $f \in C^1(\mathbb{R})$, then corresponding to each number $x_0 \in \mathbb{R}$ and each positive real number l there is a positive real number δ such that*

$$|\alpha(h_1) - \alpha(h_2)| \le l|h_1 - h_2| \ \forall h_1, h_2 \in [-\delta, \delta].$$

This follows via the mean–value theorem: one has

$$|\alpha(h_1) - \alpha(h_2)| = |\alpha'(\xi)(h_1 - h_2)|$$
$$= |f'(x_0 + \xi) - f'(x_0)||h_1 - h_2| \le l|h_1 - h_2|,$$

provided that $|h_1|, |h_2|$ are sufficiently small. $\square$

Theorem 4. *Let $f \in C^1(\mathbb{R})$ be such that $f(0) = 0$ and $f'(0) \neq 0$. There exists a number $\varepsilon_0 > 0$ with the following property: corresponding to each positive number $\varepsilon < \varepsilon_0$ there is a positive real number k such that for any function $\beta \in C^1$ satisfying $|\beta(x)|, |\beta'(x)| \le k$, the equation $f(x) + \beta(x) = 0$ has exactly one solution in the ε-neighborhood of zero. Hence, in particular, $f(x) \neq 0$ for $x \in (-\varepsilon, \varepsilon) \setminus \{0\}$.*

(Here is a somewhat imprecise but enlightening reformulation of the theorem: Given a nearly linear segment of a curve—the graph of f near the origin—cutting the x–axis only at the origin, any smooth curve–segment sufficiently close to it must likewise meet the x-axis in exactly one point.)

The proof is as follows: We have $f(x) = ax + \alpha(x)$, where $a = f'(0) \neq 0$ and $\alpha(x)/x \to 0$ as $x \to 0$. Hence the equation $f(x) + \beta(x) = 0$ may be rewritten as $x = -a^{-1}(\alpha(x) + \beta(x))$; denote by $g(x)$ the function appearing on the right-hand side of the latter equation. Let q be any fixed number in the interval $(0, \frac{1}{2})$. By Lemma 6, there exists a number $\varepsilon_0 > 0$ such that

$$|\alpha(x) - \alpha(y)| \le |a|q|x - y| \text{ for } |x|, |y| \le \varepsilon_0.$$

For each $\varepsilon < \varepsilon_0$ choose $k < \min\{\varepsilon|a|(1 - 2q), \ |a|q\}$. Then in view of the assumption $|\beta'(x)| \le k$, we have

$$|\beta(x) - \beta(y)| < |a|q|x - y|.$$

We are now ready to show that $g(x) := -a^{-1}(\alpha(x)+\beta(x))$ restricts to a contracting map from $[-\varepsilon, \varepsilon]$ to itself. For $x, y \in [-\varepsilon, \varepsilon]$, we have

$$|g(x) - g(y)| \leq |a|^{-1}(|\alpha(x) - \alpha(y)| + |\beta(x) - \beta(y)|) \leq 2q|x - y|$$

by the earlier inequalities, and for $x \in [-\varepsilon, \varepsilon]$,

$$|g(x)| \leq |g(0)| + |g(x) - g(0)| \leq |a|^{-1}k + 2q|x|$$
$$< (1 - 2q)\varepsilon + 2q\varepsilon = \varepsilon.$$

The proof is now completed by applying the fixed-point theorem for contracting maps to this restriction of g. $\square$

Exercise. Give examples showing that the conditions (a) $f'(0) \neq 0$; and (b) $|\beta'(x)| < k$ are necessary.

Corollary (The inverse function theorem). *With the same assumptions as in Theorem 4, there exist positive real numbers ε, δ and a smooth function $h : (-\delta, \delta) \to (-\varepsilon, \varepsilon)$ such that $f(h(y)) = y$ for all $y \in (-\delta, \delta)$.*

In the notation of the theorem, take $\delta := k$, and for any fixed $y \in (-\delta, \delta)$ take $\beta(x) \equiv -y$, i.e., the constant function with value $-y$. Then define $h(y)$ to be the (unique) solution of the equation $f(x) + \beta(x) = 0$, i.e., of the equation $f(x) = y$. To see that the function h is differentiable, consider the defining condition for differentiability of f at each point x_0 of the interval $(-\varepsilon, \varepsilon)$, namely $f(x) = f(x_0)+a(x_0)(x-x_0)+\alpha(x)$, $\alpha(x)/(x-x_0) \to 0$ as $x \to x_0$. Writing $y_0 := f(x_0)$ and solving for $x(= h(y))$, we obtain $h(y) = h(y_0)+a^{-1}(y-y_0-\alpha)$. Hence h will be differentiable at y_0 (with derivative a^{-1}) if $\alpha(h(y))/(y - y_0) \to 0$ as $y \to y_0$, which will follow in turn if $|h(y)| \leq c|y|$ for some positive constant c. Since this inequality is a consequence of the corollary to Theorem 3 (with the contracting map taken to be g, as defined in the proof of Theorem 4, and f the other map), we have the desired result. $\square$

Of course in the one-variable case there is a more direct proof of the inverse–function theorem; however, as noted earlier, the above proof, as also that of Theorem 4, has the virtue that it carries over without essential change in the more general situation of a function $f : \mathbb{R}^n \to \mathbb{R}^n$. In that context the best linear approximation Ax of f near the origin (assuming $f(0) = 0$) is given by taking

$$A = f' := \left(\frac{\partial f_i}{\partial x_j}(0)\right)_{i,j=1,\ldots,n},$$

the Jacobian matrix of f at 0, and the condition for invertibility of f near O (or uniqueness of solution of $Ax = 0$) becomes the invertibility of A, i.e., $\det A \neq 0$.

9.7 The Morse–Sard theorem

Exercise. Let $f \in C^1[a, b]$, and consider the set $f^{-1}(0)$ of zeros of f. Show that if $f'(x) \neq 0$ for all $x \in f^{-1}(0)$, then the set $f^{-1}(0)$ is finite.

We shall call a number x in the domain of a function f a *regular point* of f if $f'(x) \neq 0$, and a value y of f a *regular value* of f if every point in $f^{-1}(y)$ is regular (i.e., if $f(x) = y$, then $f'(x) \neq 0$).

The following theorem is an immediate consequence of Theorem 4 and the following simple fact: If a continuous function does not vanish on a closed interval, then every function sufficiently close to it is likewise nonvanishing on the interval.

Exercise. Formulate the latter assertion precisely, and prove it.

Theorem 5. *Let $f \in C^1[a, b]$ be such that $f(a)$, $f(b) \neq 0$, and suppose that 0 is a regular value of f. Then corresponding to any $\varepsilon > 0$ there is a $\delta > 0$ such that any function $g \in C^1[a, b]$ satisfying*

$$|f(x) - g(x)|, \ |f'(x) - g'(x)| < \delta \text{ for all } x \in [a, b]$$

has the same number of zeros as f in the interval $[a, b]$. Furthermore, each zero of such a function g lies in the ε–neighborhood of some zero of f.

(As mathematicians put it: The nondegenerate zeros of a class-C^1 function are stable under class-C^1 perturbations.)

Exercise. Prove that 0 is a regular value of a polynomial $p(x) = \sum_{k=0}^n a_k x^k$ if and only if all of its (real) roots are simple.

Hence if p satisfies this condition, there exists a number $\sigma > 0$ such that if $|a_i - b_i| < \sigma$, $i = 1, 2, \ldots, n$, then the polynomial $q(x) = \sum_{k=0}^n b_k x^k$ has the same number of real roots as f. However, two polynomials that seem close to one another from a naive point of view may have widely differing roots. Here is the classic example [8].

The polynomial $q(x) = \prod_{k=1}^{20}(x - k) + 2^{-23} x^{19}$ has only ten real roots, although it differs from the polynomial $p(x) = \prod_{k=1}^{20}(x - k)$, with the (simple) roots $1, 2, \ldots, 20$, only in the coefficient of x^{19} and apparently by very little: $-209.999 \cdots$ as against -210. Note, for instance, that the roots 10 and 11 of p correspond to the pair of complex conjugate roots $10.095 \cdots \pm i \cdot 0.643 \cdots$ of the polynomial q.

Since in the case of a polynomial $p(x)$ the equation $p'(x) = 0$ has only finitely many solutions, a polynomial can have only finitely many nonregular values. For an arbitrary class-C^1 function this of course need not be the case; nonetheless, in a certain sense the set of nonregular values of such a function is small. This is the significance of the following assertion, a very special case of the "Morse–Sard theorem."

Lemma 7. *The set of nonregular values of a function of smoothness class C^1 has (Lebesgue) measure zero; i.e., for $f \in C^1(\mathbb{R})$, one has*

$$\mu D = \mu\{a \mid f'(x) = 0 \text{ for some } x \in f^{-1}(a)\} = 0.$$

To appreciate this result, all one needs to know about Lebesgue measure is that a subset of the line has *measure zero* if it can be covered by a (possibly infinite) collection of intervals of arbitrarily small total length.

We begin the proof with the observation that in view of the countable additivity of Lebesgue measure, we may assume without loss of generality that the domain of f is $[0, 1]$.

Exercise. Show that the union of a countable collection of sets of measure zero again has measure zero.

Note next that since by assumption $f'(x)$ is continuous on the *closed* interval $[0, 1]$, it is in fact uniformly continuous there (this is sometimes called "Cantor's theorem"); thus given any $\varepsilon > 0$, there is a corresponding natural number n such that $|f'(x_1) - f'(x_2)| < \varepsilon$ whenever $|x_1 - x_2| < 2/n$, $x_1, x_2 \in [0, 1]$. Subdivide the interval $[0, 1]$ into n subintervals of equal length $1/n$, and let $\Delta_1, \ldots, \Delta_k$ be those of the subintervals containing a point or points at which f' vanishes. Let $x_i \in \Delta_i$ be such that $f'(x_i) = 0$. Then for any two points $y_1, y_2 \in \Delta_i$, one has

$$|f(y_1) - f(y_2)| = |f'(c)||y_1 - y_2| = |f'(c) - f'(x_i)||y_1 - y_2| < \varepsilon/n,$$

since $|c - x_i|$, $|y_1 - y_2| \leq 1/n$; hence each image set $f(\Delta_i)$ is contained in an interval of length $\leq \varepsilon/n$. We thus have that the set D of nonregular values is contained in the union $\bigcup_{i=1}^{k} f(\Delta_i)$, in turn contained in the union of a collection of $k \leq n$ intervals, each of length ε/n, and so of total length at most ε. $\square$

Note that it is not difficult to extend this proof to the more general situation of a class-C^1 map $f : \mathbb{R}^n \to \mathbb{R}^n$, where now the set D of nonregular values is defined by $D := \{f(x) \mid \det\left(\frac{\partial f_i}{\partial x_j}(x)\right) = 0\}$. On the other hand, for a function $f : \mathbb{R}^n \to \mathbb{R}^k$ with $n > k$, both the statement and proof of the theorem need to be changed. Consider, for instance, the simplest such case, namely of a function $f : \mathbb{R}^2 \to \mathbb{R}$, under the assumption that $f \in C^2$, i.e., that f has continuous second-order partial derivatives; it turns out (we shall omit the proof) that

$$\mu D = \mu\{f(x, y) \mid \tfrac{\partial f}{\partial x}(x, y) = \tfrac{\partial f}{\partial y}(x, y) = 0\} = 0.$$

We shall now use this case of the Morse–Sard theorem to investigate what happens if one changes the equation $(x^2 + 2y^2 - 1)(2x^2 + y^2 - 1) = 0$ by adding an arbitrarily small number ε to the left-hand side. (In Section 5.6 it was given as an exercise to show that the solution set of such an equation is made up of four separate pieces.) Write $f(x, y) := (x^2 + 2y^2 - 1)(2x^2 + y^2 - 1)$. Observe first that the graph of the equation $f(x, y) = 0$ consists of two ellipses intersecting at the four points $\pm(\frac{1}{\sqrt{3}}, \frac{1}{\sqrt{3}})$, $\pm(\frac{1}{\sqrt{3}}, -\frac{1}{\sqrt{3}})$, at each of which both partial derivatives of f vanish, so that 0 is a nonregular value of f. Hence we have by Morse–Sard that in particular there must exist negative regular values of f arbitrarily close to 0; i.e., there exist arbitrarily small numbers ε such that for any point (x_0, y_0) at which $f(x_0, y_0) = -\varepsilon$, the partial derivatives $\partial f/\partial x(x_0, y_0)$ and $\partial f/\partial y(x_0, y_0)$ are not both zero. Supposing, for instance, that $\partial f/\partial y(x_0, y_0) \neq 0$, we consider the map $F : \mathbb{R}^2 \to \mathbb{R}^2$ defined by $F(x, y) := (x, f(x, y) + \varepsilon)$, which at (x_0, y_0) has Jacobian matrix

$$A := \begin{pmatrix} \frac{\partial F_1}{\partial x} & \frac{\partial F_1}{\partial y} \\ \frac{\partial F_2}{\partial x} & \frac{\partial F_2}{\partial y} \end{pmatrix}(x_0, y_0) = \begin{pmatrix} 1 & 0 \\ \frac{\partial f}{\partial x}(x_0, y_0) & \frac{\partial f}{\partial y}(x_0, y_0) \end{pmatrix}.$$

Since $\det A = \partial f/\partial y(x_0, y_0) \neq 0$, we infer from the inverse–function theorem (the corollary to Theorem 4, extended to the situation of functions $\mathbb{R}^2 \to \mathbb{R}^2$) the existence of a smooth function $G : (u, v) \longmapsto (G_1(u, v), G_2(u, v))$, such that $(G_1(u, v), G_2(u, v))$ is, in some neighborhood of the point (u_0, v_0), where $u_0 = F_1(x_0, y_0) = x_0$ and $v_0 = F_2(x_0, y_0) = f(x_0, y_0) + \varepsilon = 0$, the unique solution of the system

$$x = u,$$
$$f(x, y) = v.$$

Hence in particular, the function $h(u) := G_2(u, -\varepsilon)$ is the unique solution of the system

$$x = u,$$
$$f(x, y) = -\varepsilon,$$

so that $f(u, h(u)) = -\varepsilon$; i.e., the graph of the equation $f(x, y) = -\varepsilon$ is, in some neighborhood of (x_0, y_0), that of a function. This, together with the aforementioned exercise from Section 5.6, indicates that the four crescent-shaped regions ("lunes") between the ellipses comprising the cross-section of the surface $z = f(x, y)$ by the xy-plane $z = 0$, have separated into four ovals in the lower cross-section by the plane $z = -\varepsilon$. (That there cannot be more than four ovals was established in Section 5.6.)

Exercise. What do cross-sections of the surface $z = f(x, y)$ by horizontal planes just *above* the xy-plane look like?

9.8   The law of conservation of energy

We now return to differential equations. It is of course very far from being the case that the solutions of a differential equation can always be given explicitly in terms of familiar functions; on the contrary, that possibility is rarely realizable. A compromise, basic to our further discussion, is suggested by the following lemma.

Lemma 8. *If φ is any solution of the differential equation $y'' + \sin y = 0$, then the function E defined by $E(t) := \frac{1}{2}(\varphi'(t))^2 - \cos \varphi(t)$ is constant.*

Differentiating E, one has

$$\dot{E} = \frac{d}{dt} E(t) = \varphi'(t)\varphi''(t) + \varphi'(t) \sin \varphi(t) = \varphi'(t)(\varphi''(t) + \sin \varphi(t)) = 0. \square$$

Exercise. Prove that if $\varphi''(t) + \omega^2\varphi(t) = 0$, then

$$\omega^2(\varphi(t))^2 + (\varphi'(t))^2 = \text{const.},$$

and if $\varphi''(t) - k^2\varphi(t) = 0$, then

$$-k^2(\varphi(t))^2 + (\varphi'(t))^2 = \text{const.},$$

without in either case using the explicit formulae for the solutions of these differential equations obtained earlier.

Such expressions, i.e., functions constant on any solution of a system of differential equations, are called *integrals* of the system.

Our topic in this section is that of "conservative systems with one degree of freedom" [1], meaning simply (mechanical) systems that can be described mathematically by a differential equation of the form

$$\ddot{x} = -U'(x), \; x \in \mathbb{R}.$$

The function U is called the *potential* (or *potential energy*) of the system. (We shall assume that it is as well-behaved a function as necessary.) Instead of this equation, involving the second derivative of the unknown function $x(t)$, it turns out to be useful to consider the equivalent system

$$\begin{cases} \dot{x} = y, \\ \dot{y} = -U'(x), \end{cases}$$

involving only first derivatives. The function $E : \mathbb{R}^2 \to \mathbb{R}$, defined by $E(x, y) := y^2/2 + U(x)$, is called the (total) *energy* of the system. It turns out to be an integral of the system.

Theorem 6 (Law of conservation of energy). *The function E is constant on any solution of the above conservative system; i.e., if φ, ψ are functions of t satisfying the equations of the system, then $E(\varphi(t), \psi(t)) =$ const.*

Differentiating, we have

$$\dot{E}(t) = \frac{d}{dt} E(\varphi(t), \psi(t)) = \frac{d}{dt} \left(\frac{1}{2}(\psi(t))^2 + U(\varphi(t)) \right)$$
$$= \psi(t)\psi'(t) + U'(\varphi(t))\psi(t)$$
$$= \psi(t)(-U'(\varphi(t))) + U'(\varphi(t))\psi(t) = 0. \quad \square$$

If (φ, ψ) is a solution of the system, we call the set $\{(\varphi(t), \psi(t)) \mid t \in (a, b)\} \subset \mathbb{R}^2$ a *trajectory* of the system. Recall that the *level curves* (or *level sets*, or *contours*) of the function $E(x, y)$ are essentially just the cross-sections of its graph, i.e., the sets of the form $\{(x, y) \mid E(x, y) = c\}$. In terms of these concepts the above theorem may be restated in the following geometric form: Each trajectory of a conservative system is contained in some level curve of the energy function of the system. As the following example shows, the curves of constant energy may consist of several, even infinitely many, trajectories.

The example in question is the familiar one of a pendulum, i.e., a rod fixed at one end and swinging freely in the earth's gravitational field, with a weight attached to the free end. Let m denote the mass of the pendulum (i.e., of rod and weight together), and l the distance from the point of suspension of the pendulum to its center of mass. We assume that the pendulum moves in a vertical plane, and write x for the (variable) angle it makes with the vertical (see the diagram). From Newton's second law of motion, one infers that $ml\ddot{x} = -mg\sin x$, or $\ddot{x} = -g/l \cdot \sin x$ (verify!).

Exercise. Prove that if $x(t)$ is a solution of the differential equation of the pendulum, then the function $\varphi(\tau) := x(\sqrt{l/g}\ \tau)$ satisfies $\varphi'' = -\sin\varphi$.

Hence it suffices to consider the differential equation $\ddot{x} = -\sin x$, or, equivalently, the system

$$\begin{cases} \dot{x} = y, \\ \dot{y} = -\sin x. \end{cases}$$

Thus we see that (the equation of) the frictionless pendulum constitutes a conservative system with potential $U(x) = -\cos x$.

Exercise. Show that the the the quantity $mg(U(x_1) - U(x_2))$ represents the change in potential energy of the pendulum corresponding to the change from x_1 to x_2 in the angle x.

By the law of conservation of energy, each trajectory of this system is contained in a level set of the total energy function $E(x, y) = y^2/2 - \cos x$, i.e., is part of the solution set of an equation of the form $|y| = \sqrt{2(c + \cos x)}$, $c = \text{const}$.

Clearly, if $c < -1$, the solution set is empty (whence we infer that the system cannot have energy < -1). The level set corresponding to $c = -1$ consists of the discrete set of points of the form $(2\pi k, 0)$, $k \in \mathbb{Z}$. For each value of c in $(-1, 1)$, the level set consists of a countable infinity of curves, as depicted in Figure a above. Of particular interest is the case $c = 1$, where the level set is given by $|y| = 2\cos\frac{x}{2}$. Here each point of the form $((2k + 1)\pi, 0)$ constitutes a trajectory by itself (corresponding to the unstable equilibrium with the pendulum balanced vertically above the point of suspension), and subdivides the level set into the open arcs $y = \pm\cos\frac{x}{2}$, $x \in (-\pi + 2k\pi, \pi + 2k\pi)$, each of which is a trajectory corresponding to the motion of the pendulum from a position arbitrarily close to the unstable equilibrium right around in a near circle back to the vicinity of the unstable equilibrium (see Figure b). Finally, it is clear that for energy levels higher than 1 the trajectories are circles traced out repeatedly, passing through the upper and lower equilibrium positions.

Exercise. Sketch the trajectories of a conservative system with potential (a) having graph as in Figure c above; b) given by the formula $U(x) := -\frac{k}{x} + \frac{M^2}{2x^2}$.

9.9 Small oscillations

Returning to the equations

$$\begin{cases} \dot{x} = y, \\ \dot{y} = -U'(x), \end{cases}$$

for a general conservative system, observe that if x_0 is such that $U'(x_0) = 0$, then the constant functions $x(t) = x_0$, $y(t) = 0$ afford a solution, and conversely. Thus the equilibrium points of the system lie on the x-axis, at the critical points of the potential U. We call such a critical point *nondegenerate* if $U''(x_0) \neq 0$.

It is known from physics that stable equilibrium points correspond to minimum points of the potential energy. This provides the motivation for the following mathematical definition of "stable equilibrium" for systems of differential equations such as that above. A point (x_0, y_0) is called a *stable equilibrium point* of the system if corresponding to each positive number ε there is a positive number δ such that every solution $(x(t), y(t))$ starting from a point within a distance δ of (x_0, y_0) (i.e., such that $(x(0), y(0)) \in D_\delta(x_0, y_0)$), stays within a distance ε of (x_0, y_0) at all later times (i.e., $(x(t), y(t)) \in D_\varepsilon \; \forall t \geq 0$).

Theorem 7. *If ξ is a nondegenerate minimum point of the potential function U of a conservative system, then $(\xi, 0)$ is a stable equilibrium point of the system.*

Since ξ is a nondegenerate minimum point of U, we must have $U''(\xi) > 0$, from which it follows that $U(x) > U(\xi)$ for all $x \neq \xi$ in some open neighborhood of ξ. Hence there exists a number r such that $E(x, y) = y^2/2 + U(x) > E(\xi, 0) =: c$ for all $(x, y) \in D_r(\xi, 0)$, $(x, y) \neq (\xi, 0)$. Write $\varepsilon_1 := \min\{\varepsilon, r\}$ and $\sigma := \min\{E(x, y) \mid |x|^2 + |y|^2 = \varepsilon_1^2\}$.

It is easy to see that $\sigma > c$, whence by the continuity of $E(x, y)$ there exists a number $\delta > 0$ such that $E(x, y) < \sigma$ for all $(x, y) \in D_\delta(\xi, 0)$. Hence if $(x(t), y(t))$ is any solution with initial point $(x(0), y(0)) \in D_\delta$, then $E(x(t), y(t)) = E(x(0), y(0)) < \sigma < E(x, y)$ for all $t \geq 0$, and for every point (x, y) on the boundary of the ε_1-neighborhood D_{ε_1} of $(\xi, 0)$. Thus such a trajectory must be confined to this neighborhood of $(\xi, 0)$ for all $t \geq 0$. $\square$

Knowledge of an integral (in the present case the energy integral) of a system of differential equations often allows one to obtain the solution corresponding to a particular value of the integral (here, energy level) by means of "quadrature"—i.e., integration—of explicit expressions in the elementary functions. In the situation we are concerned with, we have $\dot{x} = y = \pm\sqrt{2(E - U(x))}$, whence

$$\frac{\dot{x}}{\pm\sqrt{2(E - U(t))}} = 1.$$

Let $t_1 < t_2$ be such that $E - U(x) > 0$ on the interval between $x(t_1) =: x_1$ and $x(t_2) =: x_2$; then integrating the above equation from x_1 to x_2, we obtain

$$t_2 - t_1 = \pm\int_{x_1}^{x_2} \frac{dx}{\sqrt{2(E - U(x))}}.$$

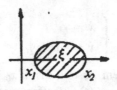

Now let ξ be a nondegenerate minimum point of the potential function U. In view of Theorem 7, the level curves corresponding to energy levels above, but close to, $E(\xi, 0)$ must be smooth curves staying in the vicinity of the point $(\xi, 0)$. Since $(\xi, 0)$ is in fact a minimum point of $E(x, y)$, it is at least intuitive that such level curves will in fact be closed curves (since one expects horizontal cross-sections of a bowl-shaped surface to be ovals); let x_1, x_2 be the x-coordinates of the points where such a periodic trajectory $(x(t), y(t))$ meets the x-axis (as in the diagram).

Since the period of such a periodic trajectory, i.e., time to complete one circuit, is twice the time taken to traverse an arc of it joining $x_1(E)$ to $x_2(E)$, we infer from the above equation the following theorem:

Theorem 8. *In a conservative system, the period of such a closed trajectory of energy level E is given by the formula*

$$T(E) = 2\int_{x_1(E)}^{x_2(E)} \frac{dx}{\sqrt{2(E - U(x))}}. \qquad \square$$

Exercise. Prove that $T(E) = \frac{dS}{dE}$, where $S = S(E)$ is the area enclosed by the closed trajectory of energy level E.

(Note first that $S(E) = 2\int_{x_1(E)}^{x_2(E)} |y| dx = 2\int_{x_1(E)}^{x_2(E)} \sqrt{2(E - U(x)}dx$, and then differentiate with respect to E using the following extended version of Leibniz's rule allowing differentiation under an integral sign: For suitably well-behaved functions $f(x, \tau), a(\tau), b(\tau)$, one has

$$\frac{d}{d\tau}\int_{a(\tau)}^{b(\tau)} f(x, \tau)dx = f(b(\tau), \tau)b'(\tau) - f(a(\tau), \tau)a'(\tau) + \int_{a(\tau)}^{b(\tau)} \frac{\partial f}{\partial \tau}(x, \tau)dx.)$$

We now investigate the trajectories of a conservative system in the vicinity of an equilibrium point $(\xi, 0)$ satisfying $U'(\xi) = 0$, $U''(\xi) \neq 0$. By moving the origin to the point $(\xi, 0)$ and changing the potential by an additive constant, we can arrange that the equilibrium point is the origin, and that $U(0) = U'(0) = 0$, $U''(0) \neq 0$. This assumed, we have

$$U(x) = \frac{1}{2}U''(0)x^2 + o(x^2),$$

and the equations of the system take the form $\dot{x} = y$, $\dot{y} = -U''(0)x + o(x)$. The linear system ("system of small oscillations" if $U''(0) > 0$)

$$\begin{cases} \dot{x} = y, \\ \dot{y} = -U''(0)x, \end{cases}$$

is thus the linear approximation to the original system near the equilibrium point $(0, 0)$.

If $U''(0) > 0$, we may write $\omega^2 := U''(0)$, and this linear system is then equivalent to the differential equation $\ddot{x} + \omega^2 x = 0$, solved in Section 4 (see Lemma 4), with its solutions periodic of period $2\pi/\omega = 2\pi/\sqrt{U''(0)}$. If on the other hand $U''(0) < 0$, then writing $k^2 := -U''(0)$, we find (see Lemma 4 again) that the trajectories are the branches of the hyperbolae $y^2 - k^2 x^2 = $ const., and rays asymptotic to these.

It is natural to expect that solutions of the approximating linear system will be close to solutions of the original system in some neighborhood of the equilibrium point $(0, 0)$. The following theorem may be regarded as confirming this.

Theorem 9. (a) *Consider a conservative system for which $U(0) = U'(0) = 0$, $U''(0) > 0$. Then the periods of trajectories with energy levels $E > 0$ approach the period of the trajectories of the corresponding system of small oscillations as $E \to 0$, i.e.,*

$$\lim_{E \to 0^+} T(E) = \frac{2\pi}{\sqrt{U''(0)}}.$$

(b) *If on the other hand $U''(0) < 0$, then the tangent lines to the branches of the level curves corresponding to the zero energy level have equations $y = \pm\sqrt{-U''(0)}\,x$, which are trajectories of the approximating linear system (see the diagram).*

Here is a sketch of the proof. For (a), note that for any small positive number E, the curve $E(x, y) = E$ may be rewritten as $y^2/2 + U''(0)x^2/2 + o(x^2) = E$, which is approximately an ellipse with semi-axes $\sqrt{2E}$ and $\sqrt{2E/U''(0)}$ and area $2\pi E/\sqrt{U''(0)}$. It follows that

$$T(E) = \frac{dS}{dE} = \frac{2\pi}{\sqrt{U''(0)}} + o(1).$$

In case (b), where $U''(0) < 0$ and $E = 0$, we have

$$y = \pm\sqrt{2(E - U(x))} = \pm\sqrt{-U''(0)x^2 + o(x^2)}$$
$$= \pm x\sqrt{-U''(0)}\sqrt{1 + o(1)}. \quad \square$$

Exercise. Prove that if $(\xi, 0)$ is a nondegenerate maximum point for the potential U, then this point represents an an unstable equilibrium.

So far in this (and in the preceding) section we have considered only systems of differential equations modeling rather idealized mechanical systems, in which, in particular, frictional forces were neglected (thus ensuring energy conservation). We now reconsider the example of the pendulum, made more realistic by introducing into the basic equation a term $a\dot{x}$, $a > 0$, in order to take account of the air-resistance to the pendulum's motion; this is appropriate, since at small speeds the air-resistance a moving object encounters is approximately proportional to the speed of the object. Thus the emended equation is $\ddot{x} + a\dot{x} + \sin x = 0$, equivalent to the system

$$\begin{cases} \dot{x} = y, \\ \dot{y} = -ay - \sin x. \end{cases}$$

Exercise. Prove that the energy function of this system is strictly decreasing along any nonstationary trajectory.

(Imitate the calculation of $\dot{E}(t)$ given in the proof of Theorem 6.)

Exercise. Verify that for this modified model of the pendulum, the stationary trajectory with the pendulum hanging down vertically is still a stable equilibrium.

We conclude the chapter with the following example. Consider a wire ring on which a bead is threaded (see the diagram below). It is assumed that the bead slides freely on the ring. Suppose that the ring is standing upright and rotating about its vertical diameter at constant angular speed Ω. It is physically intuitive that for sufficiently large rates of rotation Ω, the equilibrium state with the bead at the bottom will become unstable, and that then two new stable equilibrium positions will appear.

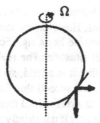

Exercise. Prove that the motion of the bead is modeled by the equations for a conservative system, with potential

$$U(x) = -\frac{g}{r}\cos x - \frac{\Omega^2}{2}\sin^2 x.$$

Sketch the level sets of the energy function and study the trajectories for various values of Ω. For which values of Ω is the equilibrium position with the bead at the bottom of the ring unstable?

Almost everything contained in this chapter is appropriate as a basis for optional courses in physics and mathematics programs in both academic and technical secondary institutions. It is necessary to state yet again the unpleasant truth that in general, highschool mathematics programs of today are suggestive of a sort of mathematical window–dressing, with an impressive-looking variety of topics given skimpy treatment. This holds true in particular for the derivative. As we well know from the history of mathematics, at the very first stage of its evolution the differential calculus was developed in response to requirements of physics; however, in modern highschool calculus courses only a nod of acknowledgment of this is given along with a few of the simplest examples. Of course, the use of the derivative in studying the behavior of functions and in graph–sketching is a worthwhile topic; however, it is *not* appropriate to limit applications to this. The author believes that the subject–matter of Sections 1 and 3 of this chapter shows clearly how to establish a connection between highschool mathematics and physics courses, and to demonstrate to the students the meaning of the words "the development of mathematics in response to the needs of physics." Apart from this, the material of those two sections is worthy of study also for methodological reasons, since in expounding the differential calculus of vector–functions based on the concept of the derivative of an ordinary (scalar) function, one is forced to review the basic definitions, properties, and results. Furthermore, the students can attempt to prove vector-analogues of the familiar theorems of ordinary calculus. A possible difficulty here arises from the need to use the machinery of vector-algebra; however since the proofs of the basic facts can be carried out in coordinatewise fashion, the teacher is given the opportunity of reviewing incidentally some analytic geometry.

A few words on the other material of this chapter: Problem 5 may be used to good effect in various ways (see, for instance, the exercise near the end of Section 2). In particular, it provides effective motivation for investigating (and solving) some rather simple differential equations. The exposition of the behavior of the solutions of the equation of a pendulum is especially noteworthy for being both relatively uncomplicated yet mathematically substantial.

In Sections 5–7, the proofs provided may sometimes seem inappropriate insofar as they concern only functions on the real line. However, the standard exposition of the differential calculus of functions of a single variable has a significant defect: most of the proofs do not extend to the multi-dimensional situation. The real line (as opposed to $\mathbb{R}^n$) has the following two peculiar properties: first it is a field, so that it seems natural to define the derivative as the limit of quotients, and second this field is ordered. The latter property has the consequence, in particular, that a real-valued function defined and continuous on an interval has an inverse function if and only if it is strictly monotonic on the interval—a result having no analogue in dimensions higher than one. The author has attempted to use strategies of proof in the one–variable situation that extend to higher dimensions; in particular, the implicit–function theorem is—almost—proved (although without being precisely stated; see the end of Section 7).

CHAPTER 10

The Foundations of Analysis

10.1   The rational and real number fields

The concepts of a natural and of a whole number (integer), or, more accurately, of the sets of natural and of whole numbers, were precisely defined in Chapter 1. The present, final, chapter is devoted to the concept fundamental to mathematical analysis, namely the set $\mathbb{R}$ of real numbers. As a preliminary to grappling with the real numbers, we give a precise definition of the rational numbers.

To this end, consider the set $\{(a, b) \mid a, b \in \mathbb{Z}, b \neq 0\}$ of ordered pairs of integers, and the relation $\sim$ on this set defined by $(a, b) \sim (c, d)$ if $ad = bc$.

Exercise. Verify that this relation is an equivalence.

Assuming this fact established, we may consider the set of equivalence classes of the relation; we denote this set by $\mathbb{Q}$, and call its elements (i.e., the classes) *rational numbers*. For the time being at least, we shall denote the class containing a pair (a, b) by $[a/b]$. Addition of rational numbers is then defined by $[a/b] + [c/d] := [(ad + bc)/bd]$, and multiplication by $[a/b] \cdot [c/d] := [ac/bd]$.

Exercise. Verify that the sum and product operations so defined are in fact well-defined (i.e., that in each case the class on the right-hand side of the defining equation is independent of the choice of the pairs $(a, b), (c, d)$ representing the classes $[a/b], [c/d]$, respectively), and that the set $\mathbb{Q}$ equipped with these operations is a field.

(Note that this construction of the rational field from the ring of integers is of a general algebraic character; it can be imitated with any integral domain in place of $\mathbb{Z}$ to obtain the "field of fractions" of that integral domain, and can in fact be extended to any commutative ring.)

A class, i.e., rational number, $[a/b]$ is said to be *positive* if $ab > 0$ and *negative* if $ab < 0$. The set of positive rationals is denoted by $\mathbb{Q}_+$. The (familiar) order relation on $\mathbb{Q}$ may then be defined as follows: $x > y$ if $x - y \in \mathbb{Q}_+$.

Recall that a field is said to be *ordered* if there is a full order defined on it that respects the two field operations in the following sense:

(1) $\forall x, y, z \ (x > y \implies x + z > y + z)$;
(2) $\forall x, y \ \forall z > 0 \ (x > y \implies xz > yz)$.

Exercise. Verify that $\mathbb{Q}$ with the above-defined order is an ordered field.

Exercise. Prove that an ordered field must have characteristic zero (see Section 5.7), and hence contains a (unique) subfield isomorphic to $\mathbb{Q}$.

Henceforth we shall use the usual notation a/b for a rational number.

It will be crucial in what follows that the field of rationals is *Archimedean*, i.e., satisfies the following "axiom of Archimedes":

$$\forall x \in \mathbb{Q} \setminus \{0\} \ \exists n \in \mathbb{N} : \ |nx| > 1.$$

(If $0 \neq x = a/b$, one may take $n = |b| + 1$.)

We are now ready to broach the theme proper of this chapter. A preliminary remark is perhaps in order: the set $\mathbb{R}$ is the first number system one encounters that as mathematical abstraction is, unlike $\mathbb{N}$, $\mathbb{Z}$, and $\mathbb{Q}$, nonalgebraic in nature.

We take the shortest—though not necessarily the easiest—route to the definition of $\mathbb{R}$, namely the axiomatic one. We call a fully ordered set on which there are given two binary operations the *field of real numbers* (the use of the definite article will be justified below) if it is an Archimedean ordered field with respect to the given order and operations, and in addition satisfies the "axiom of continuity" (or "completeness").

The latter axiom has several equivalent formulations; here are the three best known:

Cantor's version of the axiom of continuity:
   If $\{[a_n, b_n]\}_{n=1}^{\infty}$ is a nested sequence of closed intervals (i.e., $[a_{n+1}, b_{n+1}] \subseteq [a_n, b_n]$) such that $b_n - a_n \to 0$ as $n \to \infty$, then $\bigcap_{n=1}^{\infty}[a_n, b_n] \neq \emptyset$.

Cauchy's version of the axiom of continuity:
   Every Cauchy sequence has a limit.
   (A sequence $\{x_n\}_{n=1}^{\infty}$ is called *Cauchy* if

$$\forall \varepsilon > 0 \ \exists N : \ \forall m, n > N \ |x_m - x_n| < \varepsilon.)$$

Dedekind's version of the axiom of continuity:
   If A and B are two subsets such that $\forall x \in A \ \forall y \in B \ (x \leq y)$, then $\exists z : \ (\forall x \in A \ \forall y \in B \ x \leq z \leq y)$.

Theorem 1. *The above three axioms are equivalent in any Archimedean ordered field.*

Since the proofs of these equivalences are standard fare in courses in mathematical analysis, we shall not give them here. The reader may like to attempt the proofs as an exercise. □

Also standard is the proof that $\mathbb{R}$ has the "least upper bound" property that each nonempty subset S of $\mathbb{R}$ that is bounded above has a least upper bound (or "supremum"), written sup S, i.e., there is a real number a least among all upper bounds for S, in other words the least element of the set $\{b \in \mathbb{R} \mid x \leq b \; \forall x \in S\}$. In the reverse direction we have the following result:

Theorem 2. *An ordered field F with the least upper bound property must also satisfy the axiom of Archimedes and the axiom of continuity.*

The axiom of continuity is most easily deduced in Dedekind's version: just take $z := \sup A$ (in the notation of that version of the axiom).

For showing how the axiom of Archimedes follows from the least upper bound property, it is convenient first to make a definition: A nonzero element α of the field F is called *infinitesimal* if for every $n \in \mathbb{N}$ one has $|n\alpha| \leq 1$. Thus an ordered field is Archimedean precisely if it has no infinitesimal elements.

Suppose now that F is not Archimedean; the set S of infinitesimal elements is then nonempty, and the proof is completed by the following exercise.

Exercise. Prove that the set $S \subset F$ is bounded above, yet has no least upper bound. □

The field $\mathbb{R}$ is determined uniquely, i.e., completely characterized, by the above properties (or axioms); this is the sense of the following theorem.

Theorem 3. *Any two Archimedean ordered fields satisfying the axiom of continuity are isomorphic.*

For the proof we need the following lemma.

Lemma 1. *In an ordered Archimedean field F every interval contains a rational number.*

By one of the exercises above, there is a unique subfield of F isomorphic to $\mathbb{Q}$, so that we may write $F \supseteq \mathbb{Q}$. Consider any interval (a, b), $a, b \in F$, $a < b$. Since F is Archimedean, there exists a natural number n such that $1/n < b - a$. Set $m := \min\{k \in \mathbb{N} \mid k > an\}$. It is then not difficult to see that $a < m/n < b$. □

We are now ready to prove the theorem. Let $\mathbf{R}_1, \mathbf{R}_2$ be two Archimedean ordered fields satisfying the axiom of continuity, and let $\mathbf{Q}_1, \mathbf{Q}_2$ be their respective subfields isomorphic to $\mathbb{Q}$ (their "prime" subfields, as they say). Note that there is only one isomorphism between each pair of the fields $\mathbf{Q}_1, \mathbf{Q}_2, \mathbb{Q}$. For each $x \in \mathbf{R}_1$, consider the set $A_1 := \{r \in \mathbf{Q}_1 \mid r \leq x\} \subset \mathbf{Q}_1 \subset \mathbf{R}_1$, and and the image set $A_2 \subset \mathbf{Q}_2 \subset \mathbf{R}_2$. Now define a map $\varphi : \mathbf{R}_1 \to \mathbf{R}_2$, by $\varphi(x) := \sup A_2$. It is easy to see that this map is strictly monotonic (in fact, order-preserving), and so one-to-one. Surjectivity follows from the fact that $A_2 = \{r \in \mathbf{Q}_2 \mid r \leq \varphi(x)\}$.

Exercise. Verify that φ is a field isomorphism, and moreover that φ and φ^{-1} are continuous. □

In contrast with the definition of the set N of natural numbers, where essentially the only approach is the axiomatic one (i.e., via defining conditions), there are several methods for constructing the set of real numbers. Sections 4 and 5 of this chapter are devoted to these alternative ways of arriving at the reals. In the next section and Section 3 we shall investigate a rather new and unusual object: a "nonstandard" line $^*\mathbb{R}$.

10.2   Nonstandard number lines

We shall introduce these objects via axioms, i.e., defining conditions. It is not just by chance that in the heading the plural is used: unlike the axiomatic definitions of N and $\mathbb{R}$, the defining conditions for a nonstandard real line do not determine a unique such object up to isomorphism. Since these objects are rather unfamiliar, one might well ask first whether there actually exist any structures satisfying the defining conditions! In Section 10.6 we shall partially quell this doubt by constructing—or, more precisely, proving the existence of—a certain object known to satisfy the defining conditions (without, however, verifying in full that it does satisfy them).

Here then is the definition: A *nonstandard line*, denoted by $^*\mathbb{R}$, is a non-Archimedean ordered field containing $\mathbb{R}$ as a sub–ordered–field (i.e., containing a subfield isomorphic to $\mathbb{R}$ such that the restriction of the order on $^*\mathbb{R}$ to this subfield corresponds to the usual order on $\mathbb{R}$), and satisfying a further general requirement (which we postpone stating for the moment).

As we have seen in the preceding section, any such field must contain infinitesimal elements. We call an element z of such a field $^*\mathbb{R}$ *infinitely large* if $|z| > x$ for all $x \in \mathbb{R}$.

Exercise. Prove that if $\varepsilon \in {}^*\mathbb{R} \setminus \{0\}$ is infinitesimal, then $\omega := 1/\varepsilon$ is infinitely large.

As an example of a field satisfying the above explicitly stated defining conditions (leaving aside the further "general requirement" yet to be stated), one may take $\mathbb{R}(\varepsilon) := \{p(\varepsilon)/q(\varepsilon) \mid p, q \in \mathbb{R}[t], q \neq 0\}$, i.e., the field of rational functions with real coefficients, in the single variable ε. A full order is defined on this field as follows. If $p(\varepsilon) = \varepsilon^k(a_k + \cdots + a_n\varepsilon^{n-k})$, $q(\varepsilon) = \varepsilon^l(b_l + \cdots + b_m\varepsilon^{m-l})$, where $a_k, b_l \neq 0$, $k, l \geq 0$, and $n \geq k, m \geq l$, then set $p(\varepsilon)/q(\varepsilon) > 0$ if $a_k b_l > 0$. Having defined the positive elements, one then, of course, defines $p_1(\varepsilon)/q_1(\varepsilon) > p_2(\varepsilon)/q_2(\varepsilon)$ if

$$\frac{p_1(\varepsilon)}{q_1(\varepsilon)} - \frac{p_2(\varepsilon)}{q_2(\varepsilon)} = \frac{p_1(\varepsilon)q_2(\varepsilon) - p_2(\varepsilon)q_1(\varepsilon)}{q_1(\varepsilon)q_2(\varepsilon)} > 0.$$

Exercise. Verify that with this full order $\mathbb{R}(\varepsilon)$ is a ordered field.

Observe that under this ordering one has $1 - n\varepsilon > 0$ for all $n \in \mathbb{N}$, so that the indeterminate ε is an infinitesimal element of the field $\mathbb{R}(\varepsilon)$, and thence $1/\varepsilon$ infinitely large.

In the sequel we shall call the elements of $^*\mathbb{R} \setminus \mathbb{R}$ *nonstandard numbers*, or *hyperreals*, and say that two elements $u, v \in {}^*\mathbb{R}$ are *infinitely close* (written $u \approx v$) if their difference $u - v$ is infinitesimal. The elements of $^*\mathbb{R}$ that are not infinitely large will be called *finite*. For instance, in the above example both of the "numbers" $1 + \varepsilon$ and $1 + \varepsilon^2$ are finite, nonstandard, and infinitely close to 1.

Exercise. With $p, q \in \mathbb{R}[t]$ as above (in the definition of the order on $\mathbb{R}(\varepsilon)$), show that the element $p(\varepsilon)/q(\varepsilon)$ is (a) infinitely large if $k < l$; (b) infinitesimal if $k > l$; (c) infinitely close to a_k/b_l if $k = l$.

From part (c) of this exercise we see that every finite number in $\mathbb{R}(\varepsilon)$ is infinitely close to some ordinary real number. This turns out to be the case for any nonstandard line.

Theorem 4. *In any nonstandard line each finite nonstandard number u is infinitely close to a unique real number (denoted by $\mathrm{st}(u)$).*

We call $\mathrm{st}(u)$ the *standard part* of the finite nonstandard number u.

The uniqueness is easy to see, since if there existed distinct real numbers x_1, x_2 such that $u - x_1$ and $u - x_2$ were both infinitesimal, then the difference $x_1 - x_2$ would also be infinitesimal (why?), and since this number is real, we should have a contradiction.

For the existence, consider the following partition of the reals: $\mathbb{R} = L \cup R$, where $L := \{x \in \mathbb{R} \mid x < u\}$ and $R := \{x \in \mathbb{R} \mid x > u\}$. By Dedekind's version of the axiom of continuity, there exists a real number y such that $x_1 \le y \le x_2$ for all $x_1 \in L, x_2 \in R$. Suppose that the difference $u - y$ is not infinitesimal; then there exists a positive real number δ such that $|u - y| > \delta$, i.e., $u - y > \delta$ or $y - u > \delta$. If for instance $u - y > \delta$, then $y + \delta < u$, and, being real, the number $y + \delta$ therefore lies in L. On the other hand, since $y + \delta > y$, it must lie in R. The other possibility, namely $y - u > \delta$, yields a similar contradiction. $\square$

We now state the additional general condition that a nonstandard number field $^*\mathbb{R}$ is required to satisfy:

The translation principle. Every statement of ordinary real analysis should have a unique analogue over $^*\mathbb{R}$; furthermore, the original statement and its analogue should either both be true or both be false.

This principle presupposes that we are given beforehand a correspondence associating with any function $f : A \to B$, where $A, B \subset \mathbb{R}$, a unique analogue $^*f : {}^*A \to {}^*B$, where $A \subseteq {}^*A \subseteq {}^*\mathbb{R}$, $B \subseteq {}^*B \subseteq {}^*\mathbb{R}$, and $\forall x \in A \,({}^*f(x) = f(x))$, these analogues being considered the only admissible nonstandard functions in any application of the translation principle. (We shall use prefixed superscript asterisks to indicate nonstandard analogues.)

The following lemma affords examples illustrating translation.

Lemma 2. *The following statements are true:*
  (1) $f = \text{const.} \iff {}^*f = \text{const.};$
  (2) *An equation $f(x) = 0$ has a solution in $\mathbb{R}$ if and only if the equation* $^*f(x) = 0$ *has a solution in $^*\mathbb{R}$;*

(3) *The set* $^*N \setminus N$ *is nonempty, and consists of infinitely large numbers.*

For (1) we need only prove the implication $\Longrightarrow$. The assertion that the function f is constant may be expressed by

$$\forall x, y \in \mathbb{R} \ (f(x) = f(y)).$$

Hence by the translation principle, the analogous statement

$$\forall x \in {}^*\mathbb{R} \ ({}^*f(x) = {}^*f(y))$$

must also be true, i.e., $^*f = \text{const.}$

(2) As in the preceding proof, we first formulate the ordinary assertion in the language of formal logic. The statement that $f(x) = 0$ has a solution in $\mathbb{R}$ becomes

$$\exists x \in \mathbb{R} \ (f(x) = 0),$$

and by the translation principle this is true if and only if the analogous statement

$$\exists x \in {}^*\mathbb{R} \ ({}^*f(x) = 0)$$

is true.

(Clearly, this argument generalizes to systems consisting of a finite number of equations and inequalities, or, more generally, statements.)

(3) Consider the following true statement:

$$\forall x \in \mathbb{R} \ \exists n \in \mathbb{N} \ (n > x).$$

Its nonstandard analogue is

$$\forall x \in {}^*\mathbb{R} \ \exists n \in {}^*\mathbb{N} \ (n > x),$$

which must then also be true. Hence in particular, for each infinitely large number x_0 in $^*\mathbb{R}$, there is an element $\omega \in {}^*\mathbb{N}$ such that $\omega > x_0$, so that $^*\mathbb{N} \setminus \mathbb{N} \neq \emptyset$.

Exercise. Prove that $^*\mathbb{N}$ consists of positive elements of $^*\mathbb{R}$ only.

Suppose now that $\omega \in {}^*\mathbb{N} \setminus \mathbb{N}$ is finite; then there exists a natural number $k \in \mathbb{N}$ such that $k < \omega < k + 1$. However then although the system

$$\begin{cases} x \in \mathbb{N}, \\ x \in (k, k+1), \end{cases}$$

has no solutions, the element ω is a solution of its nonstandard analogue. $\square$

Exercise. Prove that $^*\mathbb{N}$ is closed under addition, and that $\forall m, n \in {}^*\mathbb{N}, m \neq n :$ $|m - n| \geq 1$.

We end this section with an example showing that in deriving the nonstandard analogues of propositions of ordinary real analysis, one must proceed with care. (It was not for nothing that in the above examples we first reformulated the original propositions in the precise language of predicate logic!)

Since the principle of mathematical induction is valid in $\mathbb{R}$, it should have a valid analogue in $^*\mathbb{R}$; hence the following proposition should be true, one might think:

$$\forall M \subseteq {}^*\mathbb{R} \; \forall x \in {}^*\mathbb{R} \; (1 \in M \; \& \; (x \in M \implies x + 1 \in M)) \implies {}^*\mathbb{N} \subseteq M.$$

However the subset $\mathbb{N}$ would seem to be a counterexample to this statement.

The fact is that in our statement of the translation principle we were a little vague: there needs to be given beforehand a prescription not just for going from an ordinary function to its nonstandard analogue, but also for going from an arbitrary subset S of $\mathbb{R}$ to its analogue ${}^*S \subseteq {}^*\mathbb{R}$. Thus only certain "intrinsic" subsets are admissible as analogues of subsets of the reals. The above paradox is then easily resolved: the set $\mathbb{N}$ cannot be an intrinsic subset of ${}^*\mathbb{R}$!

Similarly, since by the translation principle each intrinsic subset of ${}^*\mathbb{R}$ that is bounded above in ${}^*\mathbb{R}$ must have a least upper bound in ${}^*\mathbb{R}$, the subset consisting of all infinitesimal elements cannot be intrinsic.

Exercise. Prove that the set ${}^*\mathbb{N} \setminus \mathbb{N}$ is likewise not intrinsic.

10.3 "Nonstandard" statements and proofs

Exercise. Prove that ${}^*(A \cap B) = {}^*A \cap {}^*B$ and ${}^*(A \cup B) = {}^*A \cup {}^*B$.

It is, of course, not appropriate for us to attempt here a thorough exposition of "nonstandard analysis". We shall rather confine ourselves to giving a few "nonstandard" definitions, and using them to give nonstandard proofs of some familiar results of real analysis.

We first define a *nonstandard sequence* $\{x_n\}_{n \in {}^*\mathbb{N}}$ as a function from ${}^*\mathbb{N}$ to ${}^*\mathbb{R}$ that extends, via the translation principle, some function from $\mathbb{N}$ to $\mathbb{R}$, i.e., an ordinary sequence. (Thus an admissible nonstandard sequence is one extending a standard sequence in accordance with the translation principle.)

We shall need the following two technical results.

Lemma 3. (1) *For any subset S of $\mathbb{R}$, one has ${}^*S \cap \mathbb{R} = S$.*

(2) *For any infinite subset $S \subseteq \mathbb{R}$, we have that the difference set ${}^*S \setminus S$ is nonempty (and so by (1) consists of nonstandard numbers).*

(3) *A set $S \subseteq \mathbb{R}$ is bounded (above and below) if and only if ${}^*S$ consists of finite elements.*

For (1), suppose that on the contrary, ${}^*S \setminus S$ contains at least one real number s, say, and consider the identity map $\iota : S \to S$. We then have that ${}^*\iota$ is the identity map on ${}^*S$ (why?). However, then the true nonstandard assertion $\exists x \in {}^*\mathbb{R} : {}^*\iota(x) = s$ has false standard counterpart, contradicting the translation principle.

To see (2), consider any infinite sequence $\{x_n\}_{n=1}^{\infty}$ of distinct elements of S, i.e., with no two terms equal. Formally, such a sequence is given by an injection $\alpha : \mathbb{N} \to S$, which then has (unique) nonstandard extension ${}^*\alpha : {}^*\mathbb{N} \to {}^*S$ (or, less formally, the extension is a nonstandard sequence $\{x_n\}_{n \in {}^*\mathbb{N}}$, which in view of the first statement of the lemma one may think of as the original sequence with a hyperreal tail appended). Since α is an injection, ${}^*\alpha$ must likewise be injective, by the translation principle. It certainly suffices to show that for every $\omega \in {}^*\mathbb{N} \setminus \mathbb{N}$, we have ${}^*\alpha(\omega) \notin S$. Suppose that on the contrary, for some such ω, we have ${}^*\alpha(\omega) = s \in S$; then by the translation principle, since the equation ${}^*\alpha(x) = s$ has a solution in ${}^*\mathbb{N}$, its analogue $\alpha(x) = s$ must have a solution in $\mathbb{N}$, say n_0. However, then ${}^*\alpha$ sends both ω and n_0 to s, contradicting its injectivity. (More briefly: Since ${}^*\mathbb{N} \setminus \mathbb{N}$ is nonempty, the map ${}^*\alpha$ is an extension of α, and both ${}^*\alpha$ and α are injective, we must have ${}^*S \setminus S$ also nonempty.)

Proceeding to the proof of (3), we observe that if S is bounded in $\mathbb{R}$, then a direct application of the translation principle yields the same real bounds for ${}^*S$. For the converse, assume S unbounded in $\mathbb{R}$. There then exists a sequence $\{x_n\}_{n=1}^{\infty}$ of elements from S such that $\forall n \in \mathbb{N}$, $x_n > n$. Hence the analogous nonstandard sequence $\{x_n\}_{n \in {}^*\mathbb{N}}$ has the property that $\forall n \in {}^*\mathbb{N}$, $x_n > n$. Hence for any infinitely large element $\omega \in {}^*\mathbb{N}$ (i.e., any element of ${}^*\mathbb{N} \setminus \mathbb{N}$), we must have x_ω infinitely large. $\square$

Exercise. Show that for a finite subset S of $\mathbb{R}$, one has ${}^*S = S$.

Our second lemma gives a useful nonstandard reformulation of the ordinary concept of the limit of a sequence.

Lemma 4. *Let $\{x_n\}_{n=1}^{\infty}$ be an infinite sequence of reals, and let $\{x_n\}_{n \in {}^*\mathbb{N}}$ be its nonstandard analogue. Then*

$$\lim_{n \to \infty} x_n = a \iff \forall \omega \in {}^*\mathbb{N} \setminus \mathbb{N} \, (x_\omega \approx a).$$

Assuming that $\lim_{n \to \infty} x_n = a$, we have that corresponding to each real number ε there is a natural number k such that

$$\forall n \in \mathbb{N} \, (n \geq k \implies |x_n - a| < \varepsilon).$$

Hence by the translation principle, for each such pair ε, k the analogous nonstandard statement is true:

$$\forall n \in {}^*\mathbb{N} \, (n \geq k \implies |x_n - a| < \varepsilon).$$

Since each such k is real, it follows that

$$\forall n \in {}^*\mathbb{N} \setminus \mathbb{N} : |x_n - a| < \varepsilon$$

for every real positive number ε; i.e., for all $n \in {}^*\mathbb{N} \setminus \mathbb{N}$, the difference $x_n - a$ is infinitesimal.

Conversely, if $\lim_{n \to \infty} x_n \neq a$, then there exists a positive real number ε_0 such that

$$\forall k \in \mathbb{N} \, \exists n \in \mathbb{N} : n \geq k \, \& \, |x_n - a| \geq \varepsilon_0.$$

By the translation principle again, this statement with $\mathbb{N}$ replaced, wherever it occurs, by ${}^*\mathbb{N}$, must also be true; hence in particular taking k in ${}^*\mathbb{N} \setminus \mathbb{N}$ we infer the existence of an element $\omega \in {}^*\mathbb{N} \setminus \mathbb{N}$ such that $|x_\omega - a| > \varepsilon_0$, i.e., $x_\omega \not\approx a$. $\square$

We are now in a position to fulfil our promise to give nonstandard proofs of some well-known theorems of real analysis.

Theorem 5. *Any bounded, infinite subset S of $\mathbb{R}$ has a limit point in $\mathbb{R}$.*

It is not difficult to show that for every element $x \in {}^{*}S \setminus S$, the real number $y = \text{st}(x)$ is a limit point of S. $\square$

It might seem to the reader that a little more proof would be in order here; however, as the advocates of a "nonstandard" approach to teaching analysis (calculus) would say, let the traditionalists prove it.

Exercise. Prove that a function $f : S \to \mathbb{R}$ is continuous at a point $x_0 \in S$ if and only if $\forall x \in {}^{*}S \ (x \approx x_0 \implies {}^{*}f(x) \approx {}^{*}f(x_0))$.

Theorem 6. *Let $f : [a, b] \to \mathbb{R}$ be continuous. Then:*

(1) *(Weierstrass) f is bounded;*

(2) *(Cantor) f is uniformly continuous;*

(3) *(Cauchy) Every point between $f(a)$ and $f(b)$ is in the range of f (i.e., the set of values taken by f).*

For (1), observe first that ${}^{*}[a, b] = \{z \in {}^{*}\mathbb{R} \mid a \leq z \leq b\}$ (why?). In particular the elements z of ${}^{*}[a, b]$ are all finite with $\text{st}(z) \in [a, b]$. Hence by the above exercise, all values ${}^{*}f(z), z \in {}^{*}[a, b]$, of ${}^{*}f$ are likewise finite, i.e., the set ${}^{*}f({}^{*}[a, b])$ contains no infinitely large elements, whence the set $f([a, b])$ is bounded.

Exercise. Where does this argument fail for the open interval (a, b)?

For (2) consider any two infinitely close elements $x, y \in {}^{*}[a, b]$. Then $\text{st}(x) = \text{st}(y) =: c \in [a, b]$, so that ${}^{*}f(x) \approx {}^{*}f(c) \approx {}^{*}f(y)$. Hence ${}^{*}f(x) \approx {}^{*}f(y)$, which in fact shows that f is uniformly continuous on $[a, b]$.

Finally, note that (3) is equivalent to the intermediate–value theorem, so that supposing that $f(a) > 0$, $f(b) < 0$, it suffices to prove that $f(c) = 0$ for some (real) number c such that $a < c < b$. Let ω be any fixed element of ${}^{*}\mathbb{N} \setminus \mathbb{N}$, and consider the elements x_l of ${}^{*}[a, b]$ defined for each $l \in {}^{*}\mathbb{N}$ with $l \leq \omega$ by $x_l := a + l(b - a)/\omega$. There then exists an element $k < \omega, k \in {}^{*}\mathbb{N}$, such that ${}^{*}f(x_k) \geq 0$, ${}^{*}f(x_{k+1}) \leq 0$ (why?). One sees readily that $x_k \approx \text{st}(x_k) \approx x_{k+1}$, whence by the penultimate exercise again, ${}^{*}f(x_k) \approx {}^{*}f(x_{k+1})$. Hence $\text{st}({}^{*}f(x_k)) = \text{st}({}^{*}f(x_{k+1})) = f(\text{st}(x_k)) = 0$. $\square$

10.4 The reals numbers via Dedekind cuts

In this section we construct the real numbers by means of Dedekind cuts of the positive rationals [21]. We define a *cut* in the set $\mathbb{Q}_+$ of positive rationals to be a partition $L \dot\cup R = \mathbb{Q}_+$ of $\mathbb{Q}_+$ into two (disjoint) nonempty subsets L, R, with the properties that $\forall x \in L \ \forall y \in R \ (x < y)$, and that L has no largest element. (Note, incidentally, that either one of L, R by itself determines the cut; see the exercise below.) A *positive real number* is then defined to be such a cut $L \dot\cup R$; we denote the set of all of these by $\mathcal{R}$. Each positive rational r may be identified with the cut with $L := \{q \in \mathbb{Q} \mid 0 < q < r\}$ and $R := \{q \in \mathbb{Q} \mid q \geq r\}$, so that we may consider $\mathbb{Q}_+ \subset \mathcal{R}$.

Exercise. Prove that a cut in $\mathbb{Q}_+$ as defined above is completely determined by a proper, nonempty subset L of $\mathbb{Q}_+$ that is bounded above and has the following further properties:

(1) $\forall x \in L \; \exists y \in L : x < y$.

(2) If $0 < y < x$ and $x \in L$, then $y \in L$.

For any given cut, we shall call the rational numbers belonging to the left-hand set L of the cut *lower*, and those in the right-hand set R *upper*. It will be convenient also to use a single symbol, for instance ξ, to denote a cut, with the sets of lower and of upper rationals of the cut ξ denoted respectively by L_ξ and R_ξ.

We define an order relation on the set of cuts by defining $\xi \leq \eta$, for any two cuts ξ, η, if $L_\xi \subseteq L_\eta$. It will be convenient in what follows to use the following alternative characterization of this relation.

Exercise. Prove that $\xi < \eta$ if and only if there exists a rational that is a lower number for the cut η but upper for the cut ξ.

Theorem 7. *The set $\mathcal{R}$ is fully ordered by the above-defined relation.*

The reflexivity, antisymmetry, and transitivity of the relation are immediate consequences of the fact that the relation "$\subseteq$" of set–inclusion has these properties. (The map $\xi \longmapsto L_\xi$ is, evidently, one–to–one and order-preserving.) $\square$

Lemma 5. *For any two cuts ξ, η, the set $L := L_\xi + L_\eta = \{x + y \mid x \in L_\xi, y \in L_\eta\}$ is again the set of lower numbers of a cut.*

We verify that L has the properties of the penultimate exercise. If $u, v \in \mathbb{Q}_+$ are upper bounds for L_ξ, L_η respectively, then L is bounded above by $u + v$. To verify property (1) of the aforementioned exercise, observe that if $z = x + y \in L$, where $x \in L_\xi, y \in L_\eta$, then there exist rationals $x_1 \in L_\xi, y_1 \in L_\eta$ such that $x_1 > x, y_1 > y$, so that we have $z = x + y < x_1 + y_1 \in L$. Finally, to verify condition (2) of the exercise, consider $0 < w < z = x + y \in L$, with $w \in \mathbb{Q}_+$ and x, y as before. Since $0 < \frac{w}{x+y} x < x$, we have $\frac{w}{x+y} x \in L_\xi$; similarly, $\frac{w}{x+y} y \in L_\eta$. Hence $w = \frac{w}{x+y} x + \frac{w}{x+y} y \in L$. $\square$

In view of this lemma, we may define the *sum* $\xi + \eta$ of two cuts ξ, η to be the cut with set of lower elements $L_{\xi+\eta} := L_\xi + L_\eta$.

Exercise. Prove that this addition on $\mathcal{R}$ is commutative and associative.

It is easy to see that for two cuts that are (i.e., correspond to) rational numbers, their sum is just the cut corresponding to their sum as rational numbers.

Exercise. Prove that if ξ, η are cuts with $\xi < \eta$, and ζ is any cut, then $\xi + \zeta < \eta + \zeta$.

Lemma 6. *For every positive rational number r and every cut ξ, there exist numbers $x \in L_\xi, y \in R_\xi$ such that $y - x = r$.*

Let $z \in L_\xi$ be arbitrary, and consider the set $M := \{n \in \mathbb{N} \mid z + nr \in R_\xi\}$. Clearly, this set is nonempty, and therefore has a least element n_0, say. Hence $y := z + n_0 r \in R_\xi$, but $x := z + (n_0 - 1)r \notin R_\xi$. Therefore, we must have $x \in L_\xi$. Since $y - x = r$, we have the desired result. $\square$

Theorem 8. *For any cuts* ξ, η *satisfying* $\xi < \eta$, *there is a unique cut* θ *such that* $\eta = \xi + \theta$.

Consider the set $L := \{x - y \mid x > y, x \in L_\eta, y \in R_\xi\}$; we first show that this is the set of lower elements of a cut. It is not difficult to see that L is bounded above and $L \neq \emptyset$. As before, we now verify conditions (1) and (2) of the earlier exercise characterizing the set of lower elements of a cut. For (1), let $u = x - y, x \in L_\eta, y \in R_\xi$ be any element of L; there then exists an element $x_1 \in L_\eta$ such that $x_1 > x$, whence $u_1 := x_1 - y > u$, with $u_1 \in L$. To see that condition (2) holds, consider any rational w satisfying $0 < w < x - y \in L$ with x, y as before; then $0 < w + y < x$, whence $w + y \in L_\eta$, and then $(w + y) - y = w \in L$.

Thus L determines a cut θ, say. We now show that indeed $\xi + \theta = \eta$. By definition of addition of cuts, we have

$$L_{\xi+\theta} = \{z + u \mid z \in L_\xi, u \in L_\theta\}$$
$$= \{z + x - y \mid x > y, x \in L_\eta, y \in R_\xi, z \in L_\xi\}.$$

Since $z + x - y = x - (y - z) < x$, we have $z + x - y \in L_\eta$, whence $L_{\xi+\theta} \subseteq L_\eta$, i.e., $\xi + \theta \leq \eta$. For the reverse inequality, we need to show that $L_\eta \subseteq L_{\xi+\theta}$. For this it suffices to show that $L_\eta \cap R_\xi \subseteq L_{\xi+\theta}$. Let $x \in L_\eta \cap R_\xi$ (which is nonempty, since $\eta > \xi$), and let $x_1 > x$, $x_1 \in L_\eta$; then by the preceding lemma, there exist numbers $y_1 \in R_\xi$ and $z_1 \in L_\xi$ such that $y_1 - z_1 = x_1 - x$, i.e., $x = x_1 - y_1 + z_1 \in L_{\theta+\xi}$. $\square$

We call the cut θ of the theorem, the *difference* between the cuts η, ξ, and write $\theta = \eta - \xi$.

Exercise. Define an appropriate product operation on $\mathcal{R}$, i.e., a multiplication of cuts, and show that it has the usual properties.

Since $\mathcal{R}$ is, like $\mathbb{Q}_+$, fully ordered, one can imitate the construction of cuts in $\mathbb{Q}_+$ and consider cuts in $\mathcal{R}$. Among such cuts there will be those determined by elements $\xi \in \mathcal{R}$, i.e., of the form $\mathcal{L}_\xi \dot{\cup} \mathcal{R}_\xi$, where $\mathcal{L}_\xi := \{\eta \in \mathcal{R} \mid \eta < \xi\}$ and $\mathcal{R}_\xi := \{\eta \in \mathcal{R} \mid \eta \geq \xi\}$. It turns out that all cuts in $\mathcal{R}$ are of this form, so that one obtains no further entities via these cuts. (This amounts to completeness—see below.)

Theorem 9. *Every cut in* $\mathcal{R}$ *has the form* $\mathcal{L}_\theta \dot{\cup} \mathcal{R}_\theta$ *for some* $\theta \in \mathcal{R}$.

Let $\mathcal{L}$ denote the set of lower elements of an arbitrarily given cut in $\mathcal{R}$, and consider

$$L := \bigcup_{\xi \in \mathcal{L}} L_\xi.$$

Using the earlier exercise once more, one easily verifies that L is the set of lower elements of a cut θ, say, in $\mathbb{Q}_+$. (Thus for instance L has no largest element, since by the definition of a cut, no L_ξ has such an element.) Since $L_\xi \subset L$ for every $\xi \in \mathcal{L}$, one has $\xi < \theta$ for these ξ. If η is any upper element of the cut in $\mathcal{R}$ that we are considering, then since $\eta > \xi$ for all $\xi \in \mathcal{L}$, one has $L_\eta \supset L_\xi$ for these ξ, whence $L_\eta \supseteq L$, i.e., $\eta \geq \theta$. $\square$

The construction of the set $\mathbb{R}$ of all real numbers from the set $\mathcal{R}$ is completely analogous to the construction of the set $\mathbb{Z}$ from $\mathbb{N}$, described in Section 2 of Chapter 1. Note that instead of using equivalence classes, one can essentially by fiat simply take the negative reals (in Chapter 1, the negative integers) to be denoted by the positive reals with a minus sign attached, introduce zero, and posit the behavior of this enlarged set with respect to addition and the ordering; of course the remaining field axioms then have to be accommodated. We shall confine ourselves here to just one of the several verifications required before all of the familiar properties of the field $\mathbb{R}$ are finally established, namely that the rules for operating with signs with respect to multiplication follow from associativity of multiplication together with the distributive law.

Lemma 7. *In* $(\mathbb{R}, +, \cdot)$, *one has* $(-1)a = -a$, $(-a)b = -ab$, $(-a)(-b) = ab$.

Note that the first of these identities is not a tautology, in the sense that the right-hand side is *defined* to be the left-hand side, since $-a$ is normally defined rather as the additive inverse of a, i.e., as the solution of the equation $a + x = 0$. In particular, -1 is that number ω satisfying $1 + \omega = 0$. Hence $0 = (1 + \omega)a = a + \omega a$, whence $-a = \omega a = (-1)a$. The other two rules now follow easily: $(-a)b = (-1)ab = -(ab)$, and $(-a)(-b) = a(-1)(-b) = a(-(-b)) = ab$. $\square$

Theorem 10. *The fully ordered set* $(\mathbb{R}, \leq)$ *is complete.*

Exercise. Deduce this from Theorem 9 using Dedekind's version of the axiom of continuity (i.e., of the concept of completeness). $\square$

10.5   Construction of the reals via Cauchy sequences

This method of arriving at the real numbers, i.e., via Cauchy sequences of rationals, is important, since it generalizes more readily than the procedure using Dedekind cuts, and so provides a means for completing more general spaces. It also underlies the usual concrete idea of a real number as an "infinite decimal," i.e., an infinite string of integers between 0 and 9 with a decimal point placed somewhere but with no infinite "tail" of repeated 9s allowed (and with a sign appended); for example: $-5/4 = -1.25000\ldots$, $2/11 = 0.18181818\ldots$, $\sqrt{2} = 1.414\ldots$, $\pi = 3.14159\ldots$, $-e = -2.71828\ldots$. The reason for excluding infinite tails of 9s is made clear by the following exercise.

Exercise. Prove that the sequence $\{0.1, 0.19, 0.199, 0.1999, \ldots\}$ converges to 0.2.

Consider the set of all Cauchy sequences $\bar{x} = \{x_n\}$ of rational numbers (defined using rational εs only!), equipped with the following binary relation: $\bar{x} \sim \bar{x}'$ if $x_n - x'_n \to 0$ as $n \to \infty$. This, one easily verifies, is an equivalence relation, and one defines the *real numbers* to be the equivalence classes of this equivalence, thus forming in their totality the set $\mathbb{R}$. For example, by the preceding exercise, the constant sequence $\{0.2, 0.2, 0.2, \ldots\}$ is equivalent under this definition to the sequence $\{0.19, 0.199, 0.1999, \ldots\}$, and so they determine the same real number, i.e., lie in the same equivalence class. In representing the real numbers by infinite decimals, one is in effect choosing a representative from each equivalence class; e.g., π is represented by the sequence $\{3, 3.1, 3.14, 3.141, \ldots\}$ of its successive rational approximations.[1]

The following exercise establishes this.

Exercise. Consider the sequence $\{a_0, \ a_0.a_1, \ a_0.a_1a_2, \ \ldots\}$ of successive rational approximations (truncations) of an infinite decimal $\{a_0.a_1a_2a_3\ldots\}$.

(a) Prove that this sequence is Cauchy.

(b) Show that each equivalence class of Cauchy sequences contains a unique such sequence of approximations of some infinite decimal.

Each rational number q is naturally identifiable with the class consisting of all sequences of rationals converging to q, which of course contains the constant sequence $\bar{q}$ with $x_n = q$ for all $n \in \mathbb{N}$. We may therefore consider $\mathbb{Q} \subset \mathbb{R}$ (in fact, a subfield). The rationals are then just those real numbers whose decimal representations have an infinitely repeating segment; this is the point of the following exercise.

Exercise. Prove that an infinite decimal $a_0.a_1a_2a_3\ldots$ has a "periodic tail" (i.e., there exist positive integers k, l such that $a_{i+k} = a_i$ for all $i \geq l$) if and only if the corresponding equivalence class contains a constant (rational) sequence.

We shall need the following technical fact.

Lemma 8. *If $\bar{x} \neq \bar{0}$ (i.e., $\lim x_n \neq 0$), then there exists a natural number N such that either $x_n > 0$ for all $n \geq N$, or $x_n < 0$ for all $n \geq N$.*

Since $x_n \not\to 0$, there exists a (rational) number $\varepsilon > 0$ and an increasing sequence $\{n_k\}$ of natural numbers such that $|x_{n_k}| > \varepsilon$ for all $k \in \mathbb{N}$. Among the terms of the sequence $\{x_{n_k}\}$ there must, of course, be either infinitely many negative or infinitely many positive; for definiteness we suppose the latter, and then by reindexing we may suppose that in fact, the sequence $\{x_{n_k}\}$ has all its terms positive. Since the original sequence is Cauchy, there exists a natural number N such that $|x_i - x_j| < \varepsilon/2$ for all $i, j > N$. Choose $n_s \geq N$. Then for all $i \geq N$ one has $x_i = x_{n_s} + x_i - x_{n_s} \geq x_{n_s} - |x_i - x_{n_s}| > \varepsilon - \varepsilon/2 > 0.$ $\square$

---

[1] *Translator's note.* In this context how are addition and multiplication defined on $\mathbb{R}$? The answer is that term–wise addition and multiplication of Cauchy sequences of rationals determine appropriate corresponding well–defined operations on the set $\mathbb{R}$ of equivalence classes, with respect to which it is a field with the expected structure. Note that for the decimal representations of the real numbers, term–wise addition and multiplication of sequences is closely related to ordinary addition and multiplication of infinite decimals—although in performing the usual algorithms one has necessarily to work from left to right instead of the usual right–to–left.

Exercise. Calculate $1.1010010001\ldots + 0.919129123912349\ldots$.

Corollary. *If $\bar{x} \sim \bar{x}' \not\sim \bar{0}$, then there exists a natural number N such that for all $n \geq N$ the nth terms x_n and x'_n have the same sign.*

This lemma and corollary allow one to define a full–order on the set $\mathbb{R}$ of equivalence classes of Cauchy sequences of rationals (extending the usual ordering of the rationals) and to show that every such class not containing $\bar{0}$ has a multiplicative inverse.

Theorem 11. *The set $\mathbb{R}$ with the noted addition, multiplication, and full–order is an ordered Archimedean field.*

Exercise. Prove this theorem. □

In order to investigate the completeness of the field we have constructed, we need to define an appropriate metric on it. To this end, let X, X' be any two elements of $\mathbb{R}$ (i.e., equivalence classes of Cauchy sequences of rationals), and define the *distance* between these classes to be the class containing the sequence with nth term $\varepsilon := |x_n - x'_n|$, where $\{x_n\}$ and $\{x'_n\}$ are representatives of the classes X and X' respectively.

Exercise. Verify that the defining conditions for a metric are indeed fulfilled by this "distance" and that one obtains thereby a topology on $\mathbb{R}$.

Our final theorem asserts the completeness (in Cauchy's version) of the field $\mathbb{R}$ we have constructed. The reader may like to attempt the proof independently.

Theorem 12. *Every Cauchy sequence in $\mathbb{R}$ has a limit.*

10.6   Construction of a model of a nonstandard real line

A natural approach to constructing a nonstandard number field might seem to be that of declaring infinitesimals to be infinite sequences of (ordinary) reals with limit zero. However it is not difficult to see that this definition leads to nonzero elements without reciprocals (i.e., without multiplicative inverses). However, this idea is not completely wrong; it just needs to be refined considerably.

Consider the set of all infinite sequences of real numbers, equipped with the following relation: $\bar{x} := \{x_n\}_1^\infty \sim \bar{y} := \{y_n\}_1^\infty$ if $x_n = y_n$ for "almost all" $n \in \mathbb{N}$, or, more formally (and more precisely), if

$$x_n = y_n \ \forall n \in A \in \mathcal{F},$$

where $\mathcal{F}$ is a certain fixed collection of subsets of $\mathbb{N}$, i.e., a certain subset of $\mathcal{P}(\mathbb{N})$, the set of all subsets, or "power set," of $\mathbb{N}$.

Before specifying the collection $\mathcal{F}$, we shall carry out a little preliminary investigation in order to see what properties it should have. We leave the proof of the following lemma to the reader as an exercise.

Lemma 9. *The above–defined relation on the set of all infinite sequences of reals is an equivalence relation if and only if the collection $\mathcal{F}$ is nonempty and satisfies the following further conditions:*
(R1) $\forall A \subseteq B \subseteq \mathbb{R} \, (A \in \mathcal{F} \Longrightarrow B \in \mathcal{F})$;
(R2) $\forall A, B \subseteq \mathbb{R} \, (A, B \in \mathcal{F} \Longrightarrow A \cap B \in \mathcal{F})$. $\square$

In the case $\mathcal{F} = \mathcal{P}(\mathbb{N})$ (or, equivalently, $\emptyset \in \mathcal{F}$), one sees that all sequences are equivalent; we shall thus henceforth consider only nonempty collections $\mathcal{F}$ satisfying (R1) and (R2) and the further condition $\emptyset \notin \mathcal{F}$. Such a collection is called a *filter*.

Exercise. Prove that if $\mathcal{F}$ is a filter, then one cannot have for any subset $A \subseteq \mathbb{N}$ that both A and $\mathbb{N} \setminus A$ belong to $\mathcal{F}$.

The following two collections are clearly filters:

$$\mathcal{F}_0 := \{A \mid |\mathbb{N} \setminus A| < \infty\}; \quad \mathcal{F}_1 := \{A \mid 1 \in A\}.$$

However, we shall see below that neither of them suits our purpose.

Exercise. Prove that the set of equivalence classes of sequences determined by the filter $\mathcal{F}_1$ can be naturally identified with $\mathbb{R}$.

The set of equivalence classes of sequences of reals determined by a filter $\mathcal{F}$ will be denoted by $^*\mathbb{R}$. We seek the filter $\mathcal{F}$ whose corresponding $^*\mathbb{R}$ has the properties of a hyperreal field. For any filter $\mathcal{F}$, termwise addition and multiplication of sequences,

$$\{x_n\} + \{y_n\} := \{x_n + y_n\}, \text{ and } \{x_n\} \cdot \{y_n\} := \{x_n y_n\},$$

determine corresponding operations on the equivalence classes, i.e., on $^*\mathbb{R}$, turning it into an associative and commutative ring (verify!)

Lemma 10. *The ring $^*\mathbb{R}$ is a field if and only if the filter $\mathcal{F}$ determining it satisfies the following condition:*
(R3) $\forall A \subseteq \mathbb{N} \, (A \notin \mathcal{F} \Longrightarrow \mathbb{N} \setminus A \in \mathcal{F})$.

Clearly, $\{x_n\}_1^\infty \not\sim \bar{0}$ precisely if $A := \{k \mid x_k = 0\} \notin \mathcal{F}$. If $^*\mathbb{R}$ is a field, then the sequence $\{y_n\}$ with $y_k = 0$ for the same k for which $x_k = 0$ (i.e., for $k \in A$), and otherwise $y_n = 1/x_n$, must represent the class inverse to $\{x_n\}$ (why?), i.e., $\{x_n y_n\} \sim \bar{1}$. Since $x_n y_n = 1$ for $n \in \mathbb{N} \setminus A$ and is otherwise zero, it follows from the definition of the relation $\sim$ that $\mathbb{N} \setminus A \in \mathcal{F}$. Conversely, if $\mathbb{N} \setminus A \in \mathcal{F}$, then the class of the sequence $\{y_n\}$ will be a multiplicative inverse for the class containing $\{x_n\}$. $\square$

Lemma 11. *A filter satisfies the additional condition (R3) if and only if it is maximal (with respect to set inclusion) in the set of all filters.*

Exercise. Prove this assertion. $\square$

For this reason such a filter is called an *ultrafilter*. Observe that the filter $\mathcal{F}_0$ is not an ultrafilter, while $\mathcal{F}_1$ is.

A filter $\mathcal{F}$ containing a finite set must contain a smallest such set S, say, necessarily nonempty, and then $\mathcal{F}$ will consist of all subsets of $\mathbb{N}$ containing S. If $|S| > 1$, the filter is clearly not maximal. If on the other hand S is a singleton set, then $\mathcal{F}$ will be an ultrafilter, but (see the exercise preceding Lemma 10) the field $^*\mathbb{R}$ it determines is isomorphic merely to $\mathbb{R}$. Thus an ultrafilter appropriate to our purpose should contain no finite sets. Any ultrafilter containing $\mathcal{F}_0$ clearly has this property.

Theorem 13. *Every filter is contained in some ultrafilter. Hence in particular there exists an ultrafilter $\mathcal{F}^*$ containing the filter $\mathcal{F}_0$.*

It might at first seem obvious that every filter should be contained in some maximal (i.e., ultra–) filter. However, as in many situations where one would like to have a maximal element of one kind or another, as soon as one tries to construct such an ultrafilter, it begins to seem that "obvious" is not at all the right word. As a matter of fact one needs here the famous "Zorn's lemma," equivalent to the infamous Axiom of Choice of set theory. Zorn's lemma is as follows: In a partially ordered set S with the property that every ascending chain (i.e., ascending fully ordered subset) is bounded above, there exist maximal elements (i.e., there is at least one element $m \in S$ such that for all $s \in S$ one has $m \not< s$).

In the present case, we have in the role of the partially ordered set S the collection of all filters $\mathcal{F}$ containing any particular filter $\mathcal{F}'$, partially ordered by set–inclusion. The union of any ascending chain of such filters is again such a filter (verify!—or see that it almost verifies itself) and hence affords an upper bound for the chain. Thus Zorn's lemma applies, and we have the existence of the desired ultrafilter. $\square$

We shall henceforth understand $^*\mathbb{R}$ to be the field of equivalence classes of real sequences determined by $\mathcal{F}^*$.

Remark: Proofs of "mere" existence using Zorn's lemma are unsatisfying to mathematicians of the "intuitionist" persuasion, for reasons exemplified by the following consideration: The sequence $\{0, 1, 0, 1, 0, \ldots\}$ represents a class, i.e., "number," from $^*\mathbb{R}$, and clearly that number is either 1 or 0 according as the set of odd or the set of even natural numbers is an element of $\mathcal{F}^*$; however, there is, it would seem, no way of deciding which it is, since in Theorem 13 we established only the bare existence of the ultrafilter $\mathcal{F}^*$.

Note the following further property of the ultrafilter $\mathcal{F}^*$.

Exercise. Prove that if $A \cup B \in \mathcal{F}^*$, then either $A \in \mathcal{F}^*$ or $B \in \mathcal{F}^*$.

We define a full order on $^*\mathbb{R}$ as follows: Firstly, for individual sequences $\bar{x} = \{x_n\}_1^\infty, \bar{y} = \{y_n\}_1^\infty$, we define $\bar{x} \leq \bar{y}$ if the set $A := \{n \mid x_n \leq y_n\}$ belongs to $\mathcal{F}^*$. If $A \notin \mathcal{F}^*$, then $\mathbb{N} \setminus A = \{n \mid x_n > y_n\} \in \mathcal{F}$, and then we set $\bar{y} < \bar{x}$. This defines a "preorder" on the set of all sequences of reals.

Exercise. Show that this preorder determines a full order on the set $^*\mathbb{R}$.

This preorder, and the full–order on $^*\mathbb{R}$ that it determines, is somewhat unintuitive, since for instance if $\bar{x} = \{x_n\}$ and $\bar{y} = \{y_n\}$ are two sequences (with the same limit) such that $x_1 < y_1 < x_2 < y_2 < x_3 < \ldots$, then there would seem on the face of it to be no natural way of choosing between $\bar{x} < \bar{y}$ and $\bar{y} < \bar{x}$.

Exercise. Prove that the equivalence class (i.e., element of *R) containing the sequence $\{\frac{1}{n}\}$ is a positive infinitesimal in *R.

Theorem 14. *With the above–defined full-order on it, *R becomes a non-Archimedean ordered field, containing $\mathbb{R}$ as an (ordered) subfield.*

Exercise. Prove this theorem. □

Thus for the field *R to qualify as a nonstandard number field, it only remains to establish the translation principle for it. We shall now define the appropriate nonstandard analogues of standard sets of reals and functions between them, and give a partial verification of the validity of the translation principle with these as the nonstandard analogues.

Thus for any subset $M \subseteq \mathbb{R}$, we define its nonstandard analogue to be the set *$M \subseteq$ *R consisting of all equivalence classes containing sequences $\{x_n\}$ such that $\{n \mid x_n \in M\} \in \mathcal{F}^*$.

Exercise. Show that if a class contains one such sequence, then it contains only such sequences.

The proof of the following lemma is not difficult. (We leave it to the reader as an exercise.)

Lemma 12. *For any subsets $A, B \subseteq \mathbb{R}$, one has *$(A \cap B) =$ *$A \cap$*B, *$(A \cup B) =$ *$A \cup$*B.* □

The nonstandard analogue of a function $f : M \to N$, $M, N \subseteq \mathbb{R}$, is then defined to be the map *$f :$ *$M \to$ *N, where for each $u \in$ *M, the value *$f(u)$ is taken to be the class of the sequence $\{f(x_n)\}$ where $\{x_n\}$ is any representative of the class u. We leave to the reader the task of checking the validity of this definition.

The validity of the translation principle with these as the nonstandard analogues of real sets and functions would require a lengthy excursion in the forests of mathematical logic. We shall confine ourselves here to verifying that principle (in one direction) for a special type of system consisting of an equation and an inequality.

Theorem 15. *Let f, g, h, and k be real functions with respective domains F, G, H, and K (subsets of $\mathbb{R}$). If the system*

$$\begin{cases} *f(x) = *g(x), \\ *h(x) \neq *k(x), \end{cases}$$

*has a solution in *R, then the corresponding standard system*

$$\begin{cases} f(x) = g(x), \\ h(x) \neq k(x) \end{cases}$$

has a solution in $\mathbb{R}$.

Let $\{x_n\}$ be a representative of a solution (in $^{*}\mathbf{R}$) of the first system, and consider the subset A_F of $\mathbb{N}$ defined by $A_F := \{n \mid x_n \in F\}$, and the similarly defined subsets A_G, A_H, and A_K; these sets all belong to $\mathcal{F}^{*}$ (why?). Consider also $B := \{n \mid f(x_n) = g(x_n)\}$, and $C := \{n \mid h(x_n) = k(x_n)\}$; since the equivalence class of $\{x_n\}$ is a solution of the system, we must have $B \in \mathcal{F}^{*}$, $C \notin \mathcal{F}^{*}$. Thus we also have $B, \mathbb{N} \setminus C \in \mathcal{F}^{*}$. Hence

$$A_F \cap A_G \cap A_H \cap A_K \cap B \cap (\mathbb{N} \setminus C) \in \mathcal{F}^{*},$$

and then if n_0 is any number in this intersection, the term x_{n_0} will be a solution of the second system. $\square$

10.7   Norms on the rationals

In the construction of the set $\mathbb{R}$ as the completion of the rationals (in Section 5), we used implicitly a (rational) "norm" on $\mathbb{Q}$, namely the absolute value $|x|$, $x \in \mathbb{Q}$, the relevant properties of which are as follows: (1) $|x| \geq 0$, with $|x| = 0$ if and only if $x = 0$; (2) $|xy| = |x| \cdot |y|$; (3) $|x + y| \leq |x| + |y|$. It is thus natural to ask whether there are other norms on $\mathbb{Q}$, since completions (via Cauchy sequences) of $\mathbb{Q}$ with respect to these may yield exotic and interesting fields.

Let p be any fixed prime number. For each nonzero integer n, we define $\mathrm{ord}_p n$ to be the largest integer $k \geq 0$ such that p^k divides n, and then for each nonzero rational number m/n, we define $\mathrm{ord}_p m/n := \mathrm{ord}_p m - \mathrm{ord}_p n$. We now define a norm $|\ |_p$ on $\mathbb{Q}$ by setting $|x|_p := p^{-\mathrm{ord}_p x}$ for $x \neq 0$, and $|0|_p := 0$.

Lemma 13. *The function $|\ |_p$ defines a rational norm on the field $\mathbb{Q}$.*

We need to verify that (1) $|x|_p \geq 0$, and $|x|_p = 0$ precisely if $x = 0$; (2) $|xy|_p = |x|_p|y|_p$; and (3) $|x + y|_p \leq |x|_p + |y|_p$. Property (1) is obvious and property (2) is easy. Here is the verification of (3).

If x or y is zero, we obviously have equality, so we may assume that $x, y \neq 0$. Let $x = a/b$, $y = c/d$ in reduced form, $a, b, c, d \in \mathbb{Z}$. Using the fact that the largest power of p dividing the sum of two integers is greater than or equal to the largest such power dividing both summands, we have

$$\mathrm{ord}_p(x + y) = \mathrm{ord}_p \frac{ad + bc}{bd} = \mathrm{ord}_p(ad + bc) - \mathrm{ord}_p b - \mathrm{ord}_p d$$

$$\geq \min\{\mathrm{ord}_p(ad), \mathrm{ord}_p(bc)\} - \mathrm{ord}_p b - \mathrm{ord}_p d$$

$$= \min\{\mathrm{ord}_p a + \mathrm{ord}_p d, \ \mathrm{ord}_p b + \mathrm{ord}_p c\} - \mathrm{ord}_p b - \mathrm{ord}_p d$$

$$= \min\{\mathrm{ord}_p a - \mathrm{ord}_p b, \ \mathrm{ord}_p c - \mathrm{ord}_p d\} = \min\{\mathrm{ord}_p x, \ \mathrm{ord}_p y\}.$$

Hence

$$|x + y|_p = p^{-\mathrm{ord}_p(x+y)} \leq \max\{p^{-\mathrm{ord}_p x}, p^{-\mathrm{ord}_p y}\}$$

$$= \max\{|x|_p, |y|_p\} \leq |x|_p + |y|_p. \quad \square$$

We see from this proof that the norm $|\ |_p$ satisfies the "strong triangle inequality" $|x + y|_p \leq \max\{|x|_p, |y|_p\}$; such norms are called "non-Archimedean." In comparison with more familiar norms, they have some rather unusual properties. Here is one such property.

Exercise. Prove that with respect to a non-Archimedean metric on a space, the spheres are open sets of the topology defined by that metric.

For each prime p the completion of $\mathbb{Q}$ with respect to the norm $|\ |_p$ yields a field with unusual and appealing properties, called the "p-adic number field", denoted by $\mathbb{Q}_p$ [18].[2]

In view of the strangeness (and usefulness, for instance in number theory) of the p-adic number fields, it is of considerable interest to ask whether there are yet other norms on $\mathbb{Q}$. We conclude this section with the proof that there are essentially no further exotic norms.

For any fixed number $\rho \in (0, 1)$, the norm on $\mathbb{Q}$ defined by $|x|_p^\rho := \rho^{\mathrm{ord}_p x}$ determines a metric on $\mathbb{Q}$ equivalent to that determined by $|x|_p$ (where $\rho = 1/p$), so that in particular a sequence of rationals that is Cauchy with respect to any one of these norms will be Cauchy with respect to all, and hence determine the same completion of $\mathbb{Q}$. For this reason we shall, abusing notation slightly, lump such norms together under the single symbol $|\ |_p$. There are two other norms yielding two further inequivalent metrics, namely $|\ |_\infty$, the usual absolute value, and the "discrete" norm $|\ |_0$, defined by $|x|_0 = 1$ for all $x \in \mathbb{Q}$. It turns out that these account for essentially all real norms on $\mathbb{Q}$.

Theorem 16 (Ostrovskiĭ[18]).   *Every nondiscrete real norm $\|\ \|$ on the field $\mathbb{Q}$ is equivalent to $|\ |_\infty$ or to $|\ |_p$, for some prime p.*

Consider first the case that for some natural number n one has $\|n\| > 1$. Let n_0 denote the least natural number with this property, and let α be the positive real number satisfying $\|n_0\| = n_0^\alpha$. Each natural number n can be expressed uniquely in the form

$$n = a_0 + a_1 n_0 + a_2 n_0^2 + \ldots + a_s n_0^s, \ 0 \leq a_i < n_0, \ a_i \in \mathbb{Z}, \ a_s \neq 0.$$

---

[2] *Translator's note.* It can be shown that each p-adic number can be represented uniquely as an infinite string (infinite to the left) of the form

$$\ldots a_3 a_2 a_1 a_0 . b_1 b_2 \ldots b_k, \ a_i, b_j \in \{0, 1, \ldots, p - 1\};$$

thus their representations in this form are "mirror images" of the "p-ary" representations of the real numbers (e.g., binary, ternary). The algorithms for adding and multiplying such p-adic numbers then turn out to be just the "usual" ones, proceeding from right to left.
Exercise. (a) Add the 2-adic integers $\ldots 0001.0$ and $\ldots 11111.0$.
  (b) Multiply the 2-adic integers $\ldots 000011.0$ and $\ldots 010101011.0$.
  (c) Show that among such strings with $a_i, b_i \in \{0, 1, \ldots, 9\}$, i.e., the 10-adics, there are zero–divisors.

Observe that $\|a_i\| \leq 1$, since each a_i is either zero or a natural number less than n_0. Hence

$$\|n\| \leq \|a_0\| + \|a_1\|\|n_0^\alpha + \cdots + \|a_s\|\|n_0^{s\alpha} \leq 1 + n_0^\alpha + \cdots + n_0^{s\alpha}$$

$$= n_0^{s\alpha}(1 + n_0^{-\alpha} + \cdots + n_0^{-s\alpha}) \leq n^\alpha \sum_{i=0}^\infty n_0^{-\alpha i} = Cn^\alpha.$$

It follows that for every natural number N, one has $\|n\|^N = \|n^N\| \leq Cn^{\alpha N}$, whence $\|n\| \leq \sqrt[N]{C}n^\alpha$. On letting $N \to \infty$, we obtain $\|n\| \leq n^\alpha$.

We now prove the reverse inequality. Using the same representation of n in terms of powers of n_0, we have $n_0^{s+1} > n \geq n_0^s$. Hence

$$\|n\| = \|n_0^{s+1} - (n_0^{s+1} - n)\| \geq \|n_0^{s+1}\| - \|n_0^{s+1} - n\|$$

$$\geq n_0^{(s+1)\alpha} - (n_0^{s+1} - n)^\alpha \geq n_0^{(s+1)\alpha} - (n_0^{s+1} - n_0^s)^\alpha$$

$$= n_0^{(s+1)\alpha}(1 - (1 - n_0^{-1})^\alpha) \geq C'n^\alpha.$$

Since the constant C' depends only on n_0 and α, it follows by a similar argument to the earlier one that $\|n\| \geq n^\alpha$. Hence $\|n\| = n^\alpha$.

Exercise. Given that $\|n\| = n^\alpha$, as above, deduce that:
  (a) $\|x\| = x^\alpha$ for all $x \in \mathbb{Q}$.
  (b) $\|\ \|$ is a norm if and only if $\alpha \leq 1$.
  (c) This norm is equivalent to the usual absolute value $|\ |$ on $\mathbb{Q}$.

Continuing the proof, we turn to the contrary case, namely that $\|n\| \leq 1$ for all $n \in \mathbb{N}$. Set $p := \min\{n \in \mathbb{N} \mid \|n\| < 1\}$.

Exercise. Prove that p is prime.

Assuming this established, we consider any prime q different from p. If $\|q\| < 1$, then for some natural number N we shall have $\|q\|^N < \frac{1}{2}$. Since also $\|p\| < 1$, there is likewise a natural number M such that $\|p\|^M < \frac{1}{2}$. Since p^M and q^N are relatively prime, there exist integers k, l such that $kp^M + lq^N = 1$, whence

$$1 = \|1\| = \|kp^M + lq^N\| \leq \|k\| \cdot \|p\|^M + \|l\| \cdot \|q\|^N < 1.$$

We infer from this contradiction that $\|q\| = 1$ for every prime different from p. Hence for each natural number n we must have

$$\|n\| = \|p\|^{\mathrm{ord}_p n} = \rho^{\mathrm{ord}_p n}, \quad \rho \in (0, 1).$$

Exercise. Deduce that in this situation one must have in fact $\|x\| = \rho^{\mathrm{ord}_p x}$ for all $x \in \mathbb{Q} \setminus \{0\}$. $\square$

Exercise. Prove that if $|x|_p < 1$ for every prime p, then x is an integer.

In contrast with all of the earlier chapters, the material of Chapter 10 is intended mainly for internal use by the teacher, since the author finds it difficult to imagine a class of highschool students capable of grasping much of that material. However, as far as teaching the calculus to academically oriented highschool students is concerned, it is simply not possible to avoid the precise definition of the real numbers, since otherwise none of the basic results can be established! The only reasonable approach to defining the reals is the axiomatic one.[3] In the author's opinion, as far as methodology is concerned, the most natural version of the completeness axiom (axiom of continuity) for the real numbers, is that asserting the existence of a cut point. This form of the axiom is visually striking, and requires no additional apparatus of definitions. We should emphasize again that the condition that there exist a least upper bound for each nonempty subset of real numbers that is bounded above is equivalent to completeness together with the Archimedean axiom. As for the latter axiom, the author is not at all sure that it should be stated explicitly. If one were in all seriousness to present to an unprepared class as an *axiom* the statement that

$$\forall x > 0 \; \exists n \in \mathbb{N} : 0 < \frac{1}{n} < x$$

(to which any student might respond with "Just take $n = [1/x] + 1$!"), the result would be total incomprehension.

---

[3] *Translator's note.* Defining the reals as "infinite decimals" (see Section 10.5) would seem to be a good alternative; it has at least the virtue of greater concreteness. However, the translator's experience suggests that in highschool the reals are not normally defined precisely at all (nor, for that matter, in ordinary—as opposed to honors—lower-level university courses in mathematics). Instead, the students usually just have indicated to them the one-to-one correspondence between the real numbers and the points of the line—which might at a stretch be considered as hinting at Dedekind's definition—with perhaps a bare mention of the least upper bound property.

References

[1] Arnol'd, V.I., *Mathematical Methods of Classical Mechanics*, Second Edition, Springer–Verlag, 1989. (Translated by K. Vogtmann and A. Weinstein.)

[2] Arnol'd, V.I., *Ordinary Differential Equations*, Springer–Verlag, 1992. (Translated by Roger Cooke.)

[3] Coxeter, H.S.M., *Introduction to Geometry*, Wiley, 1961.

[*4] *Encyclopaedia of Elementary Mathematics*, Vols. 1 and 2, Moscow – Leningrad, 1951.

[*5] Faddeev, D.K., *Lectures on Algebra*, Moscow, 1984.

[6] Fomin, D., Kirichenko, A., *Leningrad–Petersburg Mathematical Olympiads 1987–1991*, Series: Contests in Mathematics, Vol. 1, MathPro Press, Westford, MA, 1994. (ISBN: 0–9626401–4–X)

[7] Fomin, D., Genkin, S., and Itenberg, I., *Mathematical Circles (Russian Experience)*, Series: Mathematical World, Vol. 7, Amer. Math. Soc., 1996. (Translated by Mark Saul.) (ISBN: 0–8218–0430–8)

[8] Forsythe, G.E., Malcolm, M.A., and Moler, C.B., *Computer Methods for Mathematical Computations*, Prentice–Hall, 1977.

[9] Griffiths, H.B. and Hilton, P.J., *A Comprehensive Textbook of Classical Mathematics: A Contemporary Interpretation*, van Nostrand, London, 1970.

[*10] Gusev, V.A., Orlov, A.I., and Rozental', A.L., *General Mathematical Assignments for Grades 9–12*, Moscow, 1984.

[11] Halmos, P.R., *Lectures on Ergodic Theory*, Math. Soc. Japan, Tokyo, 1956.

[12] Hardy, G.E., Littlewood, D.E., and Pólya, G., *Inequalities*, Second Edition, Cambridge Univ. Press, 1988.

---

[0]*In Russian.

[13] Herstein, I.N., *Topics in Algebra*, Second Edition, Wiley, New York, 1975.

[14] Hilbert, D., *The Foundations of Geometry*, Second Edition, Open Court, LaSalle, 1971. (Translated from the German by Leo Unger.)

[15] Hilbert, D., and Cohn–Vossen, S., *Geometry and the Imagination*, Chelsea Publishing Co., New York, 1952. (Translated from the German by P. Neményi.)

[16] Husemoller, D., *Fibre Bundles*, McGraw–Hill, 1966.

[17] Klein, F., *Elementary Mathematics from an Advanced Standpoint*, Springer-Verlag, 1924–25. (Reprinted by Dover.)

[18] Koblitz, N., *p-adic Numbers, p-adic Analysis, and Zeta Functions*, Graduate Text in Mathematics 58, Springer-Verlag, 1977.

[19] Kostrikin, A.I., *Introduction to Algebra*, Springer-Verlag, 1982. (Translated by Neal Koblitz.)

[20] Krantz, S.G., *Techniques of Problem-Solving*, Amer. Math. Soc., 1997.

[21] Landau, E., *Foundations of Analysis. The Arithmetic of Whole, Rational, Irrational, and Complex Numbers*, Chelsea Publishing Co., New York, 1951. (Translated from the German by F. Steinhardt.)

[22] Machover, M., and Hirschfeld, J., *Lectures on Non-standard Analysis*, Springer-Verlag, 1969.

[23] Milnor, J.W., *Topology from the Differentiable Viewpoint*, Univ. Press of Virginia, Charlottesville, 1965.

[24] Milnor, J.W. *Differential Topology*, Lecture Notes, Princeton Univ., 1958.

[25] Milnor, J.W., and Husemoller, D., *Symmetric Bilinear Forms*, Springer-Verlag, 1973.

[26] Pólya, G., *Mathematics and Plausible Reasoning*, Vols. 1 & 2, Princeton Univ. Press, Princeton, 1954.

[27] Postnikov, M.M., *Foundations of Galois Theory*, Pergamon Press, Oxford, 1962. (Translated by Ann Swinfen.)

[28] Shafaravich, I.P., *Basic Algebraic Geometry*, Springer-Verlag, 1974. (Translated by K.A. Hirsch.)

[29] Shilov, G.E., *Mathematical Analysis: A Special Course*, Pergamon Press, Oxford, 1965. (Translated by J.D. Davis.)

[*30] Sominskiĭ, I.S., Golovina, L.I., and Yaglom, I.M., *On Mathematical Induction*, Moscow, 1967.

[31] Stewart, Ian, *From Here to Infinity*, Oxford Univ. Press, Oxford, 1996.

[32] Stewart, Ian, *Galois Theory*, Chapman and Hall, New York, 1973.

[33] Stillwell, J., *Mathematics and its History*, Springer-Verlag, 1994.

[*34] Sushkevich, A.K., *The Fundamentals of Higher Algebra*, Moscow–Leningrad, 1937.

[*35] Uspenskiĭ, V.A., *What Nonstandard Analysis Is*, Nauka, Moscow, 1987.

[36] Vilenkin, N.Ya., *Combinatorics*, Academic Press, New York, 1971. (Translated by Abe Shenitzer.)

[37] Wilson, R.J., *Introduction to Graph Theory*, Oliver and Boyd, Edinburgh, 1972.

[38] Yaglom, I.M., *Geometric Transformations*, Random House, New York, 1962. (Translated by Allen Shields.)

Index